A Focus on Addition and Subtraction

This innovative text offers a unique approach to making mathematics education research on addition, subtraction, and number concepts readily accessible and understandable to pre-service and in-service teachers of grades K–3.

Revealing students' thought processes with extensive annotated samples of student work and vignettes characteristic of teachers' experiences, this book provides educators with the knowledge and tools needed to modify their lessons and improve student learning of additive reasoning in the primary grades. Based on research gathered in the Ongoing Assessment Project (OGAP), this engaging, easy-to-use resource features practical resources such as:

- A close focus on student work, including 150+ annotated pieces of student work, to help teachers improve their ability to recognize, assess, and monitor their students' errors and misconceptions, as well as their developing conceptual understanding;
- A focus on the *OGAP Addition, Subtraction, and Base Ten Number Progressions,* based on research conducted with hundreds of teachers and thousands of pieces of student work;
- In-chapter sections on how Common Core State Standards for Math (CCSSM) are supported by math education research;
- End-of-chapter questions to allow teachers to analyze student thinking and consider instructional strategies for their own students;
- Instructional links to help teachers relate concepts from each chapter to their own instructional materials and programs;
- An accompanying eResource, available online, offers an answer key to Looking Back questions, as well as a copy of the *OGAP Additive Framework* and the *OGAP Number Line Continuum.*

A Focus on Addition and Subtraction marks the fourth installment of the popular *A Focus on...* collection, designed to aid the professional development of pre-service and in-service mathematics teachers. Following from previous volumes on ratios and proportions, multiplication and division, and fractions, this newest addition is designed to bridge the gap between what math education researchers know and what teachers need to know in order to better understand evidence in student work and make effective instructional decisions.

Caroline B. Ebby is a Senior Researcher at the Graduate School of Education, University of Pennsylvania.

Elizabeth T. Hulbert is Managing Partner and Professional Development Coordinator at the Ongoing Assessment Project.

Rachel M. Broadhead is Project Director for the Alabama Math, Science, and Technology Initiative, University of South Alabama.

Studies in Mathematical Thinking and Learning
Alan H. Schoenfeld, Series Editor

A Focus on Addition and Subtraction

Bringing Mathematics Education
Research to the Classroom

Caroline B. Ebby, Elizabeth T. Hulbert,
and Rachel M. Broadhead

Routledge
Taylor & Francis Group

NEW YORK AND LONDON

First published 2021
by Routledge
52 Vanderbilt Avenue, New York, NY 10017

and by Routledge
2 Park Square, Milton Park, Abingdon, Oxon, OX14 4RN

Routledge is an imprint of the Taylor & Francis Group, an informa business

© 2021 Taylor & Francis

The right of Caroline B. Ebby, Elizabeth T. Hulbert, and Rachel M. Broadhead to be identified as authors of this work has been asserted by them in accordance with sections 77 and 78 of the Copyright, Designs and Patents Act 1988.

Library of Congress Cataloging-in-Publication Data
Names: Ebby, Caroline B., author. | Hulbert, Elizabeth T., author. | Broadhead, Rachel M., 1971- author.
Title: A focus on addition and subtraction : bringing mathematics education research to the classroom / Caroline B. Ebby, Elizabeth T. Hulbert, and Rachel M. Broadhead.
Description: New York, NY : Routledge, 2021. |
Series: Studies in mathematical thinking and learning | Includes bibliographical references and index.
Identifiers: LCCN 2020023841 | ISBN 9780367481636 (hardback) | ISBN 9780367462888 (paperback) | ISBN 9781003038337 (ebook)
Subjects: LCSH: Addition–Study and teaching (Elementary) | Subtraction–Study and teaching (Elementary) | Number concept–Study and teaching (Elementary) | Mathematics teachers–Training of. | Elementary school teachers–Training of.
Classification: LCC QA135.6 .E235 2021 | DDC 372.7072–dc23
LC record available at https://lccn.loc.gov/2020023841

ISBN: 978-0-367-48163-6 (hbk)
ISBN: 978-0-367-46288-8 (pbk)
ISBN: 978-1-003-03833-7 (ebk)

Typeset in Minion
by River Editorial Ltd, Devon, UK

Visit the eResources: www.routledge.com/9780367462888

This book is dedicated to the thousands of teachers who have participated in the Ongoing Assessment Project development over the years. The willingness of educators to share suggestions and student responses has had tremendous impact on our work. Additionally, this book is dedicated to Marge Petit who has provided tireless leadership throughout the 17-year span of the Ongoing Assessment Project. Marge's vision, guidance, and commitment has benefitted OGAP immeasurably.

Contents

Preface

The Importance of Additive Reasoning

Additive reasoning is the first stage in reasoning mathematically and the focus of mathematics instruction in the primary grades. The transition from counting and early number understanding to addition, subtraction, unitizing, and base-ten understanding lays a foundation for later multiplicative, fractional, and proportional reasoning. Readers familiar with the existing OGAP learning progressions will recognize that the first level of multiplicative reasoning is characterized by strategies based on counting and addition and that these additive strategies can be built upon to develop multiplicative reasoning.

While many adults may think that the math students learn in the earliest grades is simple, that could not be further from the truth. The skills and concepts of additive reasoning are complex and provide foundational knowledge for the mathematics students will learn throughout their school years, as well as in daily life. Without a strong core of understanding of these mathematical ideas about number, addition, and subtraction, students may struggle to make sense of math for years to come. In fact, early competence with number has proven to be a predictor of future success in mathematics and of schooling in general into the high school years (Denton & West, 2002; Duncan et al., 2007; National Mathematics Advisory Panel, 2008).

A Focus on Addition and Subtraction: Bringing Mathematics Education Research to the Classroom

A Focus on Addition and Subtraction: Bringing Mathematics Education Research to the Classroom is the fourth book in a series focused on the development of important mathematics content and formative assessment anchored in the use of student work to make instructional decisions. The graphic below represents the idea that additive reasoning is at the core of mathematical reasoning, as well as a foundation for the rest of the *Focus On . . .* series. All four books in the series build on each other to provide a comprehensive roadmap for developing mathematical reasoning in grades K–8, just as all four content areas build on each other to create mathematical proficiency. Additive reasoning lays the foundation for multiplicative reasoning; multiplicative reasoning lays the foundation for fractions; fractions lays the foundation for proportional reasoning; and all of these are important for algebraic reasoning in the secondary school years. The success of the earlier books on fractions and multiplicative reasoning for grades 3 to 5 encouraged us to write similar books on both addition and subtraction and proportional reasoning. The complete series now covers the core content areas in grades K through 8.

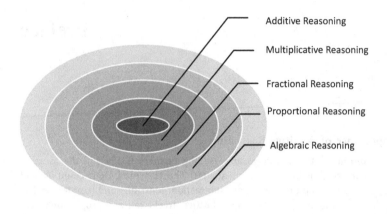

Additive Reasoning

Multiplicative Reasoning

Fractional Reasoning

Proportional Reasoning

Algebraic Reasoning

This book has its origins in a series of professional development sessions developed to provide teachers with the essential understanding of mathematical content, research, formative assessment strategies, and tools and resources to meet the needs of all students. We have found that when teachers have a deeper knowledge of the skills and concepts a student must possess to be mathematically proficient, as well as an understanding of how skills and concept build upon each other and common student struggles, they can design and deliver more effective instruction to deepen students' mathematical knowledge.

This work is part of the Ongoing Assessment Project (OGAP), which began in 2003 with a grant from the National Science Foundation, with the goal of developing research-based formative assessment tools and strategies for teachers of mathematics. The initial grant focused on grades 3–8, and over time there was a demand for this same work in the earlier grades to set a strong foundation. This book is the result of a multi-year partnership with the Consortium for Policy Research at the University of Pennsylvania and additional funding from the National Science Foundation to develop, pilot, and field-test resources and routines for grades K–3.

Central Features of this Book

Student thinking is at the center of this book. The chapters are filled with examples of authentic student work to exemplify the mathematics education research as well as our experiences in working with students and teachers in schools across the country. Much of what we have learned about the teaching and learning of early mathematics is grounded in our own experiences with teachers and students. Throughout the years of this project, we have piloted and collected thousands of pieces of student work; the work we have selected to use in the book is a typical representation of the range of responses we have gathered or observed over the years. There is text surrounding each piece of student work in order to dig deeper into the evidence of student thinking.

The *OGAP Additive Reasoning Framework* is another central component of the book. The framework was developed through the distillation of research, ongoing implementation and feedback with hundreds of teachers, and the analysis of thousands of examples of student work. It is designed to represent and communicate the mathematics education research on how students make sense of the skills and concepts related to number and base-ten understanding, addition, and subtraction at the earliest grade levels. The *OGAP Additive Reasoning Framework* contains three learning progressions: Addition, Subtraction, and Base Ten. Each progression is designed at a grain size to be usable by classroom teachers for examining and using math program materials, selecting formative assessment tasks, responding to evidence in student work, and making instructional decisions. The framework is introduced in Chapter 2 and then revisited throughout the book as we examine student work and discuss students' developing understandings and misconceptions.

Learning progressions, or learning trajectories as they are often called in mathematics education research, have had an influence on current standards and mathematics curriculum materials. The use of learning progressions in professional development and classroom instruction has also been found to impact both teacher and student learning and motivation (Carpenter, Fennema, Peterson, Chiang, & Loef, 1989; Clements, Sarama, Spitler, Lange, & Wolfe, 2011; Supovitz, Ebby, Remillard, & Nathenson, 2018). The idea that children's understanding should be frequently assessed in relation to a learning progression in order to inform and target instruction is central to this book. A recent study by Clements and colleagues (2020) showed that young learners who receive instruction based on a learning trajectory for addition and subtraction made greater gains than students who were taught at the higher levels of the trajectory. In other words, skipping levels on the learning trajectory for instruction was not beneficial for learning.

Most chapters in this book include a brief discussion of the grade level expectations of the Common Core State Standards in Mathematics (CCSSM) in relation to the mathematics focus of the chapter and the related mathematics education research. In addition, many chapters contain instructional tips and possible questions educators might ask their students to probe for understanding. These formative assessment probes are noted with a small icon:

At the end of each chapter is a section titled Looking Back which includes questions to help teachers think more deeply about some important topics that were discussed in the chapter. Many of the questions in Looking Back include examples of student work to help connect these concepts to classroom implementation. The answers to these questions can be found at www.routledge.com/9780367462888. Most chapters also include a section titled Instructional Link, which comes right after Looking Back. The questions in this section are designed to encourage teachers to reflect on and analyze their own instructional or program materials based on information within the chapter, with the goal of improving teaching and learning.

A Book for Teachers

A Focus on Addition and Subtraction: Bringing Mathematics Education Research to the Classroom is written for classroom and preservice teachers. It is our hope that classroom teachers, special educators, math interventionists, and math coaches will benefit by the depth at which we focus on the most important content of K–3 mathematics. All aspects of this book: the content, authentic student work, connections to the mathematics education research, Looking Back, and Instructional Links, should provide focused opportunities to discuss the challenges we all face as educators when helping students become mathematically proficient. Additionally, for preservice teachers this book offers an introduction to the important math content of K–3, illustrated with authentic student work, providing access to student thinking and ways of knowing you might not otherwise have opportunity to consider. Ultimately, the information in this book is meant to serve as a resource and promote thought-provoking professional dialogue towards the goal of improving teaching and learning in math classrooms everywhere.

Acknowledgments

We would like to extend our most sincere appreciation to the thousands of teachers across the country who have participated in the Ongoing Assessment Project (OGAP) Additive Reasoning professional development and piloted and shared student work with us. Teachers from Vermont, Pennsylvania, New Hampshire, Alabama, and South Carolina, as well as many other locations, have greatly influenced the work with their enthusiastic willingness to implement these ideas with their students and provide us with thoughtful feedback. We also want to thank the original OGAP design team whose early development work in other content areas provided the model for this work, as well as the members of the OGAP National Professional Development Team.

Additionally, we want to thank the late Dr Karen King from the National Science Foundation and Dr Jonathan Supovitz, director of the Consortium for Policy Research on Education (CPRE) at the University of Pennsylvania for championing our work and providing us with ongoing support and guidance over the years.

We also extend our thanks to Nicole Fletcher for her work on the project, particularly the item bank and piloting, Bridget Goldhahn for graphic design of the *OGAP Additive Framework*, Chris Cunningham and Ellie Ebby for the artwork, Siling Guo for assistance with the manuscript preparation, and Brittany Hess for her many contributions to the project.

The work presented in this book was supported by the National Science Foundation (DRL—1620888). Any opinions, findings and conclusions or recommendations expressed in this material are those of the author(s) and do not necessarily reflect those of the National Science Foundation.

1
Additive Reasoning and Number Sense

Big Ideas

- Additive reasoning includes various mathematical skills, concepts, and abilities that contribute to number sense.
- Additive reasoning is built on concepts of early number, including part–whole relationships, commutativity, and the inverse relationship between addition and subtraction.
- Additive reasoning both depends upon, and contributes to, the development of base-ten number understanding.
- The *OGAP Additive Framework* contains learning progressions that provide instructional guidance for teachers so that all students can access the important concepts and strategies that lead to additive reasoning.

Additive reasoning involves much more than being able to add and subtract. It involves knowing when to use addition and subtraction in a variety of situations, choosing flexibly among different models and strategies, using reasoning to explain and justify one's approach, having a variety of strategies and algorithms for multi-digit addition and subtraction, and knowing if an answer or result is reasonable. Moreover, it is important for fluency with addition and subtraction procedures to be built upon conceptual understanding and reasoning. As the National Council of Teachers of Mathematics (NCTM) states:

> Procedural fluency is a critical component of mathematical proficiency. Procedural fluency is the ability to apply procedures accurately, efficiently, and flexibly; to transfer procedures to different problems and contexts; to build or modify procedures from other procedures; and to recognize when one strategy or procedure is more appropriate to apply than another.
>
> (2014a, p. 1)

Additive reasoning is the focus of K–2 mathematics and provides a foundation for multiplicative reasoning in the intermediate grades. According to Ching and Nunes (2017), additive reasoning is one of the crucial components of mathematical competence and is built on conceptual understanding of number and part–whole relationships. As students learn to work with larger quantities, additive reasoning also involves understanding of the base-ten number system and relative magnitude.

Additive Reasoning: The Mathematical Foundations

Additive reasoning centers around part–whole relationships. As Van de Walle and colleagues (2014) state, "to conceptualize a number as being made up of two or more parts is the most important relationship that can be developed about numbers" (p. 136). At first students will use their counting skills to construct an understanding of the relationships between quantities, but over time they will develop strategies and concepts that move them towards reasoning additively. Understanding quantities in terms of part–whole relationships is a significant achievement which allows children to compose and decompose numbers and use those relationships to make sense of and solve problems in a range of situations. The part–whole relationship between two quantities involves understanding both the commutative property and the inverse relationship between addition and subtraction (Ching & Nunes, 2017).

Commutativity

Students often learn about commutativity when they are developing fluency with single digit addition facts. This property of addition has its roots in understanding that a quantity can be separated into two or more smaller quantities and that the order in which they are added does not change the value. Any number c can be made up of part a and part b ($c = a + b$) or part b and part a ($c = b + a$). Figure 1.1 is an illustration of both part–whole relationships and commutativity.

Figure 1.1 Nine apples broken into two groups can be thought of as 6 and 3 or 3 and 6 but in both cases, there are 9 apples

Modeling commutativity in concrete situations allows for students to be exposed to this idea while they are developing part–whole understanding. Using models to observe and generalize commutativity can be significantly more powerful for students than learning it abstractly as a rule to remember (e.g., 3 + 6 is the same as 6 + 3 because they are "turn-around" facts).

 Chapter 5 Visual Models to Support Additive Reasoning for more on the importance of visual models and Chapter 6 Developing Whole Number Addition for more on the commutative property.

The Inverse Relationship between Addition and Subtraction

The second property of addition that is central to part–whole relations, and therefore additive reasoning, is the inverse relationship between addition and subtraction. Instruction about the relationship between addition and subtraction often includes a focus on students generating a set of related equations (sometimes called "fact families"), but the concept is much more dynamic. Let's consider the 9 apples again. The inverse relationship means that taking away one part from the whole leaves the other part, so if you remove the 3 you are left with 6 apples. Furthermore, since the two parts, 3 and 6, are interchangeable parts, if you remove the 6 you are left with 3 apples. There are therefore four related equations that can represent the part–whole relationship as shown in Figure 1.2.

Figure 1.2 Four related equations

$3 + 6 = 9$

$6 + 3 = 9$

$9 - 3 = 6$

$9 - 6 = 3$

The equations in Figure 1.2 are the way we numerically represent the relationships, but having a strong understanding of the inverse relationship between addition and subtraction is the result of working with concrete objects, visual models, and various problem situations. In other words, understanding the relationships between the quantities should be central to instruction rather than simply teaching students how to write the related equations.

The integration of part–whole, commutativity, and the inverse relationship between addition and subtraction leads to the development of additive reasoning, characterized by the ability to think about the relations between the quantities when solving problems. For example, children with an understanding of the inverse relationship between addition and subtraction can solve problems modeled by equations such as $7 + x = 10$ or $x - 5 = 8$ by using the inverse operation.

Connecting Additive Reasoning and Base-Ten Understanding

According to several researchers (Krebs, Squire, & Bryant, 2003; Martins-Mourão & Cowan, 1998; Nunes & Bryant, 1996), concepts of additive reasoning must be in place in order to develop an understanding of base ten. Since our number system is composed of place value parts in varying unit sizes that combine to make the whole, flexibly working with multi-digit numbers involves concepts of part–whole, commutativity, and the inverse relationship between addition and subtraction. For example, 68 can be thought of as additively composed of 60 and 8 or 8 and 60, and if 8 is taken from 68 then 60 remains. These ideas are foundational to flexible use of the base-ten number system and number sense. At the same time, as students develop base-ten understanding they are able to develop more sophisticated strategies for addition and subtraction.

When students truly understand and can meaningfully combine these ideas, they can apply them to construct a relational understanding of numbers and operations, which in turn leads to strong number sense. The *OGAP Additive Framework*, discussed in more detail in Chapter 2, includes progressions for base ten, addition, and subtraction, as these concepts develop concurrently throughout the early elementary years.

What Is Number Sense?

Number sense is a widely used term encompassing a range of skills and concepts across all levels of mathematics. Broadly, number sense can be thought of as a flexible understanding of numbers and their relationships. According to NCTM (2000):

> As students work with numbers, they gradually develop flexibility in thinking about numbers, which is a hallmark of number sense... Number sense develops as students understand the size of numbers, develop multiple ways of thinking about and representing numbers, use numbers as referents, and develop accurate perceptions about the effects of operating on numbers.
>
> (p. 80)

Number sense does not boil down to a single skill or concept. Many of the important components that make up number sense have their origins in the earliest grades: equality, base-ten understanding, relative magnitude, operations, number relationships, and estimation, to name a few. Building students' number sense involves making connections between these concepts with a focus on understanding and flexible use to solve problems.

Additive Reasoning and Number Sense: From a Teaching and Learning Perspective

Teachers face many challenges in supporting the development of students' additive reasoning and number sense. Most teachers did not experience math instruction that was focused on developing number sense, either as a learner through their K–12 instruction or in their teacher preparation programs. An important goal is to view the learning of math as a more dynamic process, one that involves curiosity and flexibility about numbers and relationships. Number sense is a way of thinking that should permeate all aspects of math teaching and learning (Berch, 2005; Sousa, 2008), but many math program materials do not provide comprehensive tools and resources to continually develop and deepen number sense. Overcoming these challenges requires teachers to become math learners themselves, increasing the likelihood of seeing opportunities within curriculum materials to build number sense into math instruction.

Although number sense may be difficult to define, teachers often say, "you know it when you see it." Let's consider what additive reasoning and number sense look like in the primary grades. Examine the four pieces of student work shown in Figures 1.3 to 1.6. These are responses from one student to a variety of tasks given in a second grade classroom over a few months. As you look over the student work, try to make sense of the student's thinking and then think about how you would complete this sentence:

The primary grade student who has number sense shows evidence of...

Figure 1.3 Marissa solves a problem involving groups of ten

Ms. Luo's class went apple picking. The class picked 327 apples. The class put their apples in bags to take them home. If each bag can hold 10 apples, what is the fewest number of bags they need to hold their apples?

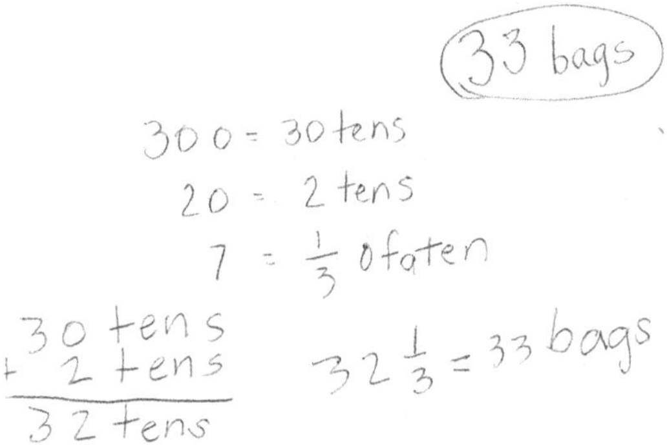

$$300 = 30 \text{ tens}$$
$$20 = 2 \text{ tens}$$
$$7 = \frac{1}{3} \text{ of a ten}$$

$$\begin{array}{r} 30 \text{ tens} \\ + \ 2 \text{ tens} \\ \hline 32 \text{ tens} \end{array}$$

$$32\frac{1}{3} = 33 \text{ bags}$$

(33 bags)

Figure 1.4 Marissa uses her understanding of the relationship between the quantities to explain why the solutions will be the same

Assata needed to solve 100 – 36. In order to solve the problem, she solved 99-35 and got the correct answer. Why does her strategy work?

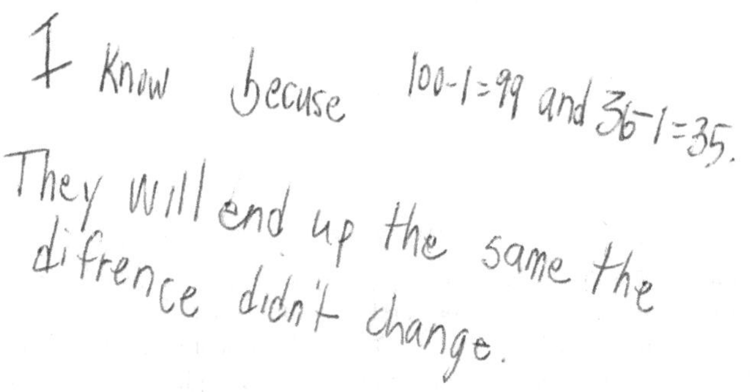

I know becuse 100-1=99 and 36-1=35. They will end up the same the difrence didn't change.

Figure 1.5 Marissa uses addition to solve a problem with a missing subtrahend

Kids baked 765 cupcakes for the school bake sale. They sold some cupcakes and 398 were left. How many cupcakes did they sell?

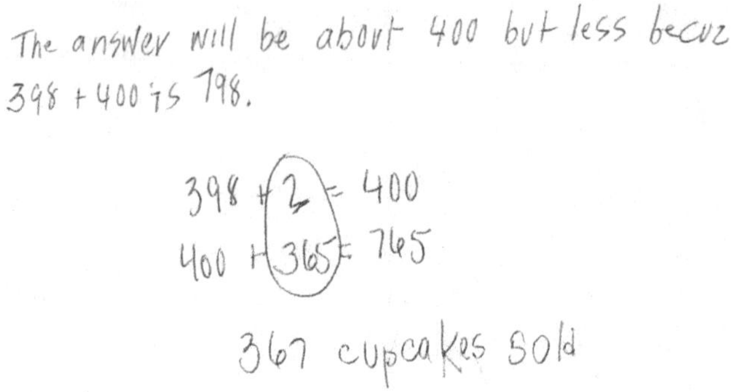

The answer will be about 400 but less becuz
398 + 400 is 798.

398 + ②= 400
400 + ③⑤⑤= 765

367 cupcakes sold

Figure 1.6 Marissa uses subtraction to solve a problem with a missing starting number

Amir had a penny collection. He found 37 more pennies. Now he has 121 pennies in his collection. How many pennies did Amir have in his collection to start with?

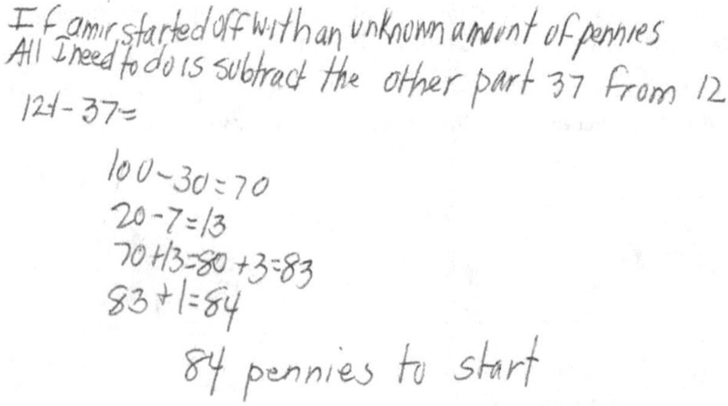

If amir started off with an unknown amount of pennies
All I need to do is subtract the other part 37 from 121.
121 - 37 =

100 - 30 = 70
20 - 7 = 13
70 + 13 = 80 + 3 = 83
83 + 1 = 84

84 pennies to start

When looking across these four pieces of student work, there is evidence that Marissa has strong understanding of number and operations and the relationships between quantities in the problems. In Figures 1.3, 1.5, and 1.6, she draws on different strategies and operations to solve each of the tasks; her strategies also reflect solid understanding of the base-ten number system and are efficient and accurate. For example, in Figure 1.3, her work shows evidence that she can decompose 327 into place value parts and then into tens, and although she makes an error about the value of 7 in the context of the problem, she is able to make sense of the situation in order to determine that she needs another bag. In Figure 1.4 she explains that decreasing each number in the original subtraction problem by one will result in the same answer or difference, demonstrating an understanding of the operation of subtraction.

In Figures 1.5 and 1.6 there is evidence that Marissa understands and uses the relationship between addition and subtraction and that she can use her base-ten understanding to construct meaningful strategies to perform the operations. In Figure 1.5 she first estimates before solving the task, providing evidence that she considers reasonableness and relative magnitude when solving problems.

Based on the student work in Figures 1.3 to 1.6, you may have come up with all or many of the characteristics below in response to the statement "The primary grade student who has number sense shows evidence of . . ."

- base-ten understanding,
- ability to decompose and recompose numbers,
- ability to estimate,
- ability to use a variety of strategies to add and subtract,
- understanding the inverse relationship between addition and subtraction,
- awareness of relationships among numbers in addition and subtraction,
- ability to explain and communicate one's thinking,
- basic fact fluency.

What other skills and concepts would you add to the list based on your own knowledge of students who demonstrate characteristics of strong number sense? Some other important skills and concepts that will be discussed in this book include:

- understanding properties of addition,
- the ability to conceptualize a collection as a group (unitizing),
- using and interacting with a variety of models to solve problems,
- using mental math strategies,
- using benchmark numbers and relative magnitude.

This is just the beginning of a long list of skills and concepts necessary to build strong additive reasoning and number sense. Developing deep and flexible number sense is a complex and time-intensive endeavor that spans many years of mathematics learning.

The *OGAP Additive Framework* and Learning Progressions

The *Ongoing Assessment Project (OGAP) Additive Framework* can be downloaded at **www.routledge.com/9780367462888**. It is a tool for educators that represents the mathematics education research on how students develop early number, base-ten understanding, and procedural fluency with addition and subtraction based on conceptual understanding, as well as common errors students make or preconceptions that interfere with learning new concepts and solving problems.

The *OGAP Additive Framework* is made up of three progressions for base-ten number, addition, and subtraction. The *OGAP Base Ten Number Progression* represents the development of early number concepts and the path to base-ten fluency. The other progressions do the same for addition and subtraction. All three progressions represent common student strategies from least to most sophisticated, moving from bottom to top. Examine the progressions on pages 2–4 of the *OGAP Additive Framework* and compare them to your list of characteristics of number sense. Where do you see those characteristics represented? Do you see other characteristics on the progressions that are not on your list? In Chapter 2 we will look more closely at the *Addition* and *Base Ten Number Progressions*.

Chapter 2 for more on the *OGAP Additive Framework*.

Developmental progressions characterize common pathways that students take when learning concepts, but students often learn at their own pace and acquire skills and concepts at different times. When teachers are aware of the skills and concepts students need to move forward, they can use that information to more effectively respond to student needs while at the same time building on their current level of understanding to move towards the mathematics. Sarama and Clements (2009) state that learning progressions can help teachers see "themselves not as moving through a curriculum, but as helping students move through levels of understanding" (p. 17) and can positively effect student motivation and achievement. There is a growing body of research to suggest that professional development in learning progressions increases teacher knowledge, instructional practices, and student learning (e.g., Clements, Sarama, Wolfe, & Spitler, 2013; Supovitz et al., 2018; Wilson, Sztajn, Edgington, & Myers, 2015).

The *OGAP Additive Framework* is a tool that represents learning progressions in a way that is useable for teachers in making more informed instructional decisions. Knowing the progression of skills and concepts that build towards additive reasoning helps teachers to be better users of their math program materials, respond to individual student needs, and continually and effectively support student learning.

The *OGAP Progressions* illustrate how mathematical strategies develop and grow in sophistication, helping teachers to appreciate that all strategies are not equal, yet also allowing access for individual students to use strategies that make sense to them. This in turn provides a more equitable approach for math instruction, allowing students to engage in the mathematics at a level appropriate for their needs and

Figure 1.7 Jana's response. Jana drew 28 marks on the line labeling each one and then drew 25 more marks and counted them all

Hasan played tag for 25 minutes and then played basketball for 28 minutes. How many minutes did Hasan play altogether?

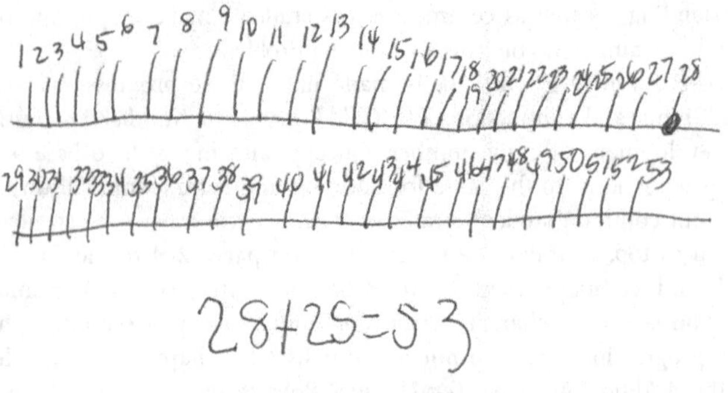

$$28 + 25 = 53$$

Figure 1.8 Zeb's response. Zeb draws base-ten blocks to solve the problem, modeling both numbers and counting by tens and then by ones

Hasan played tag for 25 minutes and then played basketball for 28 minutes. How many minutes did Hasan play altogether?

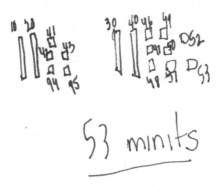

Figure 1.9 Aadi's response. Aadi decomposes 28 into 25 and 3, adds the 25s, and then the 3

Hasan played tag for 25 minutes and then played basketball for 28 minutes. How many minutes did Hasan play altogether?

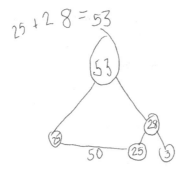

at the same time supporting their continual progress. To illustrate this idea, examine the three pieces of work from different students in the same classroom in Figures 1.7 to 1.9. How are these responses to the same task similar, and how are they different?

You likely noticed that these three students all solved the task correctly, arriving at an answer of 53, but used very different strategies. Jana's strategy shows evidence that she needed to model and count by ones to get her answer. Zeb's work shows evidence of also drawing to solve the problem, but he used units of tens and one and then counted those units to arrive at the answer. Aadi's response shows a more sophisticated strategy: he decomposed one number into useful parts for the task and then added up the parts. His visual model represents both the decomposition of the 28 and the relationship that all the parts make up the total of 53. These correct student solutions illustrate very different approaches to solving the problem and exemplify the idea that all strategies are not equal in sophistication.

In most classrooms, teachers are faced with the kind of range in students' math understanding reflected in the examples in Figures 1.7 to 1.9. Having the tools, strategies, and knowledge to move the math forward for all students in the class is an important goal of understanding and using the *OGAP Additive Framework*. Throughout this book we will use the *OGAP Addition, Subtraction,* and *Base Ten Number Progressions* to examine strategies in student work and think about possible instructional decisions based on the evidence in the work.

Ultimately our goal as educators of mathematics is to move students towards more efficient and sophisticated strategies, while at the same time assuring that they can access and build on their less efficient strategies when the mathematics gets more difficult or unfamiliar. Explicitly making connections to link students' strategies increases the likelihood that students will become powerful additive reasoners with strong number sense and many resources in their mathematical toolbelt. The chapters in this book will help teachers learn to understand and use the *OGAP Addition, Subtraction,* and *Base Ten Number Progressions* to help themselves and their students understand the progression of strategies as well as the connection between strategies.

Chapter Summary

This chapter focused on:

- The mathematical foundations of additive reasoning, including part–whole understanding, commutativity, and the inverse relationship between addition and subtraction.
- The concurrent and mutually reinforcing relationship between the development of additive reasoning and base-ten understanding.
- Characteristics of additive reasoning and number sense.
- The power of learning progressions as tools for making instructional decisions.
- An introduction to the *OGAP Addition, Subtraction,* and *Base Ten Number Progressions.*

Looking Back

1. **What Is Additive Reasoning?** Imagine being asked to speak at a parent night at the beginning of the school year. You want parents to understand the major mathematics work in primary grades and your goals for math instruction. How would you describe additive reasoning and number sense to parents?

2. **Mutually Reinforcing Ideas:** Understanding numbers in terms of part–whole relationships involves an integrated understanding of commutativity and the inverse relationship between addition and subtraction. These ideas are foundational to additive reasoning, which in turn supports base-ten understanding. Use the quantities 56, 75, and 131 to answer the following questions.

   (a) Draw a visual model to show the part–whole relationship between these three quantities.

(b) What four addition or subtraction equations can you write from your model?

(c) How is commutativity of addition involved in the four equations you wrote?

(d) How is the inverse relationship between addition and subtraction involved in the four equations you wrote?

(e) What role does base-ten understanding play in the ability to add and subtract these quantities?

3. **Learning Progressions as Tools for Instructional Decision-Making:** Students move at their own pace through developmental progressions as they learn mathematics, acquiring skills and concepts at different times. Explore the *OGAP Additive Framework* as you consider the following:

(a) In what ways does information in the progressions support the formulation of next instructional steps as you move students toward acquiring important mathematical understanding?

(b) How can the information presented in the framework inform decisions about the use curriculum materials?

4. **Making Sense of the *OGAP Addition Progression*:** Together the *OGAP Base Ten Number, Addition,* and *Subtraction Progressions* communicate the progression of skills and concepts students acquire as they become additive reasoners. Examine the *OGAP Addition Progression*.

(a) How would you describe the main differences in strategies as you move through the levels?

(b) What are the characteristics of strategies at the *Additive* level of the progression?

2
The *OGAP Additive Framework*

Big Ideas

- The *OGAP Additive Framework* is based on mathematics education research on how students develop addition, subtraction, and base-ten number fluency with understanding.
- The *OGAP Additive Framework* is designed to be a tool for teachers to use as they gather evidence of student thinking to inform instruction and monitor student learning.
- Knowledge and use of learning progressions has been shown to positively impact teachers' knowledge and instruction and students' motivation and achievement.

The *OGAP Additive Framework* was developed from mathematics education research on how students learn addition, subtraction, and number concepts and is a valuable tool to help teachers select and design tasks, examine and understand evidence in student work, make instructional decisions, and provide actionable feedback to students. This chapter provides an overview of the *OGAP Additive Framework*, which includes the *Addition Progression, Subtraction Progression*, and the *Base Ten Number Progression*. The *Base Ten Number Progression* will be introduced in this chapter and then discussed again in more detail in Chapters 3 and 4. Because the *Addition* and *Subtraction Progressions* are closely related, this chapter will introduce the main features through a focus on addition; both progressions are discussed in more detail in Chapter 6 and Chapter 7.

It is suggested that you download the *OGAP Additive Framework* and refer to it as you read this chapter and references to the framework in other chapters throughout the book. An electronic copy can be found at **www.routledge.com/9780367462888**.

There are three major elements of the *OGAP Additive Framework*:

1. *Problem Structures* (front page) that includes characteristics that may impact the difficulty of tasks and strategies students use.
2. The *OGAP Addition and Subtraction Progressions* (inside pages) that show evidence of student work along a continuum of student understanding for addition and subtraction.

3. *The OGAP Base Ten Number Progression* (back page) that shows evidence of student work along a continuum of student understanding for early number and base-ten concepts.

The parts of the *OGAP Additive Framework* are interrelated. Movement of student strategies along the progressions for addition and subtraction is often influenced and related to understanding of number (as reflected on the *Base Ten Number Progression*), the structures of the problem, and/or the problem situation. Student work is used throughout this chapter to describe and illustrate the different strategy levels on all three progressions. Instructional strategies to help develop student understanding and move their mathematical thinking along the progressions will be included in this chapter and throughout the book. *Problem Structures* will be discussed in Chapter 8.

All the progressions in the *OGAP Additive Framework* are similar in format. They are meant to reflect levels of student thinking and do not include every possible concept or strategy children need to acquire; rather they focus on the bigger ideas. They are designed to illustrate multiple pathways and instructional options depending on student understanding. In other words, there is not just one path up the progressions.

·Ultimately, the progressions included in the *OGAP Additive Framework* have been designed to serve as a resource to teachers throughout all aspects of their math instruction. Research indicates that teacher knowledge and use of learning progressions positively affects instructional decision-making and student achievement in mathematics (Carpenter, Fennema, Peterson, Chiang, & Loef, 1989; Clements, Sarama, Spitler, Lange, & Wolfe, 2011; Supovitz, Ebby, Remillard, & Nathenson, 2018).

The *OGAP Base Ten Number Progression*

The *OGAP Base Ten Number Progression* is designed to be used by teachers to gather evidence of student thinking and strategies related to counting, early number, and the development of base-ten understanding. Additionally, there is an opportunity to identify underlying issues and errors that students exhibit when solving tasks related to number development that may get in the way of transitioning to more flexible and sophisticated understanding. All the progressions provide some instructional guidance about how to transition students from one level to the next, based on the evidence in the student work.

The *OGAP Base Ten Number Progression* is designed to communicate the development of both students' understanding of quantity and relative magnitude as they interact with more complex numbers and develop more flexible thinking towards base-ten understanding. The levels on the progression provide a structure for teachers to consider a progression from early development of number and counting all the way to flexible use of base-ten understanding to solve a variety of mathematical problems and situations. The detail within each level, as well as across the levels, is at a grain size that makes the progression usable by teachers and includes instructional guidance to move students' strategies along the progression as they acquire deeper understanding and are challenged by more complex tasks. Importantly, the progressions are not meant to assess the overall level of student thinking, but rather to assess the strategies students use to solve particular problems. As students are exposed to new concepts or interact with larger numbers or unfamiliar problem situations, their strategies may move up and down the progression. In the following sections, each level on the progression is discussed and illustrated with examples of student work.

Precounting Strategies

Children learn to count before they have developed an understanding of number or quantity (i.e., the meaning of counting). As a result, at the early stages, they will often demonstrate incomplete or partial understandings when counting or comparing quantities. For example, they may not yet understand the need to tag every object with one number word, or they may not yet connect the act of counting with the result—determining the total amount. When asked to compare two collections, young children may pay more attention to how much space the collections take up than the quantity. These strategies reflect that important number concepts, discussed in detail in Chapter 3, are still developing.

In Figure 2.1 Seth was asked to tell which row has more shapes in it. His teacher noted that he pointed to the first collection saying, "it is longest." Seth's response is evidence of a *Precounting* strategy as he is comparing based on his perception of length or space rather than focusing on quantity.

Figure 2.1 *Precounting* strategy. Seth's response "it is longest" indicates evidence of perceptual comparing

Who has more shapes? Show or explain how you know.

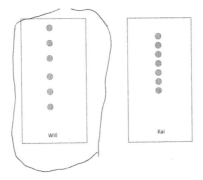

Seth's response: "It is longest."

The *Precounting* level on the *Base Ten Number Progression* illustrates some of the common errors students make or perceptual strategies they use in counting and comparing that indicate a still-developing understanding of number. This level is unlike the others on the progression which are focused on early levels of base-ten understanding rather than errors.

Early Counting Strategies

Once students develop a strong understanding of early counting, they can use counting to solve problems. When students first engage in solving tasks involving determining quantity or comparing quantities, they often treat quantities as collections of ones. Even when the configuration of the collection presents an opportunity to use grouping, students employing an *Early Counting* strategy will count by ones. In Figure 2.2 Morgan's work is evidence of an *Early Counting* strategy. She counts all the rocks in both collections by ones, and then uses the total to correctly conclude that Mia has more rocks. She does not use the groups of five that are in the illustrations to solve the task.

Figure 2.2 *Early Counting* strategy. Morgan counts all of the rocks by ones to determine who has more rocks

Mia has a rock collection and so does Jordan.

Who has more rocks?

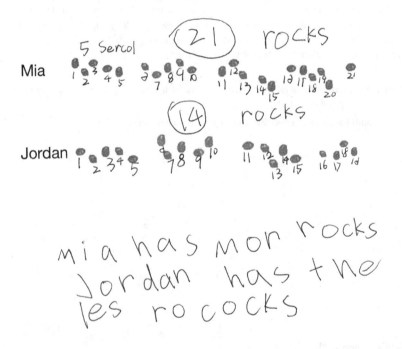

The student work in Figure 2.3 is an example of a student counting by ones to locate 25 on the number line when 0 and 50 are given. Levi's work indicates that he needs to see and use the ones in order to find the relative location of 25 on the number line. He is able to recognize and use a reasonable distance for each unit of one indicating he understands the relationship between distance and quantity on the number line.

Figure 2.3 *Early Counting* strategy. Levi uses counts by ones to place 25 on the number line

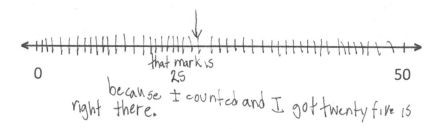

0   that mark is   50
                25
   because I counted and I got twenty five is
right there.

From Early Counting to Early Unitizing

Unitizing is the ability to see a group, call it one group, but also know that it is worth another value. As students begin to develop an ability to unitize small groups of numbers, they can move to more efficient strategies to determine quantity and compare numbers. For example, one five-frame that is filled could be thought of as one group of five or five ones. Unitizing is discussed in detail in Chapters 3 and 4 as it is an important underlying concept for base-ten understanding.

As students are able to employ unitizing to solve tasks, they can transition to *Early Unitizing* strategies. An important aspect of student thinking at this stage is the need to see the smaller units that make up a larger unit. Consuelo's response in Figure 2.4 is evidence of using her understanding of grouping by fives to locate 19 on the number line. One can observe in the work that Consuelo needs to see the ones but also shows evidence that she recognizes the units of five within her ones by writing the multiples of 5 larger than the other numbers.

Figure 2.4 *Early Unitizing* strategy. Consuelo counts by ones, indicating the units of 5s to locate 19

Where would 19 go on the number line?

From Early Unitizing to Unitizing by Composite Units

As students move towards deeper understanding of base ten and more efficient strategies, we see them working with units that are made up of smaller units, without needing to see the smaller units. This is a particularly important understanding as students begin to extend their base-ten reasoning to working with larger numbers. The transition to using larger units without needing to see or sketch the smaller units is illustrated in Figures 2.5 and 2.6.

In Figure 2.5 Ellie's work is evidence of a student who needs to see the ones to make groups of tens, resulting in a significant amount of work to arrive at a solution. While Ellie's strategy results in a correct answer, the time and effort required to get there is less efficient and unwieldy as numbers get larger. In Figure 2.6 Ben works with tens without needing to see the ones. The evidence indicates that he understands that one pack is equivalent to ten bracelets, and he

Figure 2.5 *Early Unitizing* strategy. Ellie draws every bracelet in order to put them in units of 10, counting by tens as she accounts for a bracelet for each student

Shonda bought packs of glow bracelets to give out to all the kids in her grade. Each pack has 10 glow bracelets in it. There are 56 kids in Shonda's grade. How many packs of bracelets did she need to buy?

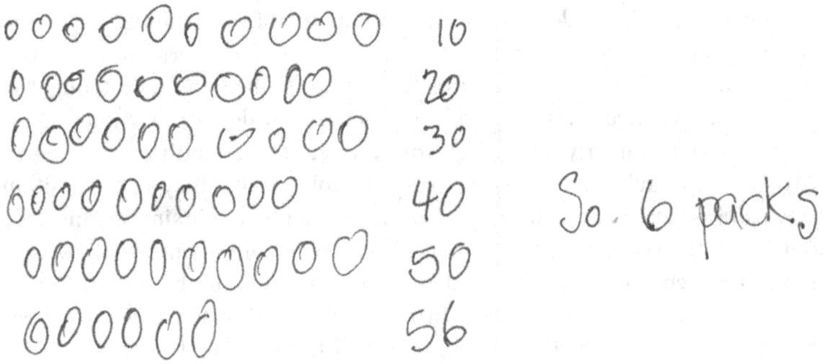

Figure 2.6 *Unitizing by Composite Units*. Ben counts by tens to figure out how many packs of bracelets are needed for the class

Shonda bought packs of glow bracelets to give out to all the kids in her grade. Each pack has 10 glow bracelets in it. There are 56 kids in the Shonda's grade. How many packs of bracelets did she need to buy?

$$10 + 10 + 10 + 10 + 10 + 6 = 56$$

Shonda needs 5 packs and 6 more bracelets so 6 packs in all.

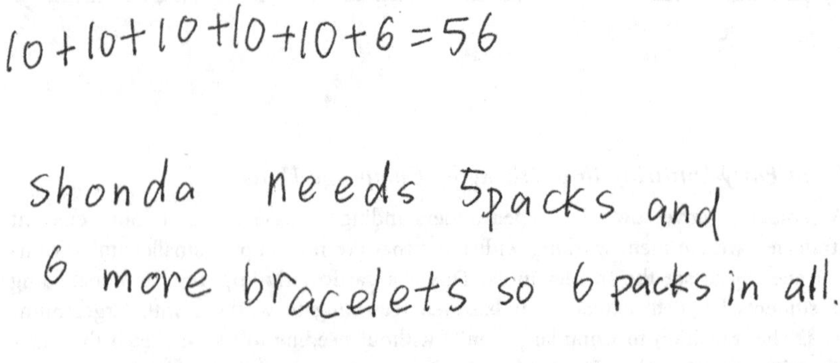

also understands that an additional pack is needed to account for the six additional bracelets.

Both students' solutions are correct, but Ellie's work indicates she needs more exposure to unitizing without seeing the smaller unit before working with larger numbers, an important distinction between strategies at the *Early Unitizing* and *Unitizing by Composite Units* levels. This is an example of how the *OGAP Base Ten Number Progression* can provide instructional guidance to move students towards deeper understanding and more efficient strategies for working with multi-digit numbers.

From Unitizing to Number Composition by Place Value Parts

At the next level on the progression, the evidence in student work indicates an understanding of the structure of numbers and the use of that understanding to decompose and recompose quantities in place value parts (ones, tens, hundreds, etc.). For example, a student could understand that 235 can be thought of as 2 hundreds, 3 tens, and 5 ones and use that knowledge to determine that 235 is also equivalent to 23 tens and 5 ones. This kind of flexible understanding of base ten can be particularly valuable as a student could compose and decompose numbers in the format most helpful for a given context or situation. They can also use their understanding of base ten and place value parts to compare quantities and consider relative magnitude.

In Figure 2.7, DeShawn uses his base-ten understanding to solve a task involving determining the number of groups of ten in a three-digit quantity. His work shows evidence of understanding the structure of the number 327, demonstrating that 10 bags hold 100 apples so 30 bags is equivalent to 300 apples. He then states that 2 bags will have 20 apples and the extra 7 apples will need another bag.

Figure 2.7 *Number Composition* strategy. DeShawn uses base-ten understanding to break apart the number 327 into groups of tens and then recomposes the tens to arrive at an answer of 33

Ms. Luo's class went apple picking. The class picked 327 apples. The class put their apples in bags to take them home. If each bag can hold 10 apples, what is the fewest number of bags they need to hold all the apples?

Figure 2.8 is an example of a student who uses place value and understanding of the relationships between numbers to consider the relative magnitude of a quantity. Becca's work indicates that she understands 50 is halfway between 0 and 100 and that this relationship will be useful for locating 66 on the number line. Then she places 66 on the number line in relation to 50 by making a jump of 10, 5, and 1 from 50.

Figure 2.8 *Number Composition* strategy. Becca's strategy of using the benchmark number 50 to locate 66 on a number line

Where does 66 go on the number line?

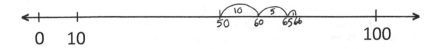

From Number Composition to Application of Base-Ten

At the top level of the progression, the evidence in student work indicates that students can use flexible and efficient base-ten strategies in a variety of situations. Through an understanding of the structure of the base-ten number system, students can move seamlessly between symbolic notation of the numeral and multiple ways of decomposing and composing the number in place value parts to solve a variety of tasks. A student with this understanding understands the connection between the base-ten system, the underlying place value of each digit, and the ten-to-one relationship between the units. It is important that this kind of flexible understanding be built on understanding and not simply memorized steps so that it can be applied to computation, estimation, and algebraic thinking. Base-ten understanding at this level goes beyond knowledge of procedures and can be readily used in other contexts, such as for efficient and flexible computation or extension to decimals and negative numbers.

In Figure 2.9 Marta shows evidence of understanding the structure of the base-ten number system by explaining how the 33 bags are related to the quantity of 327 apples without needing to decompose the 327 into hundreds and then tens. She also indicates that she knows 3 more apples would be needed to fill the last bag of ten.

As will be discussed throughout this book, the flexible understanding of the base-ten structure of the number system, illustrated at the top level of the progression, is built on deep conceptual understanding and is the goal for instruction in grades K–3. In turn, students' understanding of number and base ten impacts their ability to develop more efficient strategies for addition and subtraction, as discussed in the next section.

Figure 2.9 *Application of Base Ten.* Marta uses her flexible understanding of base ten to solve the task

Ms. Luo's class went apple picking. The class picked 327 apples. The class put their apples in bags to take them home. If each bag can hold 10 apples, what is the fewest number of bags they need to hold all the apples?

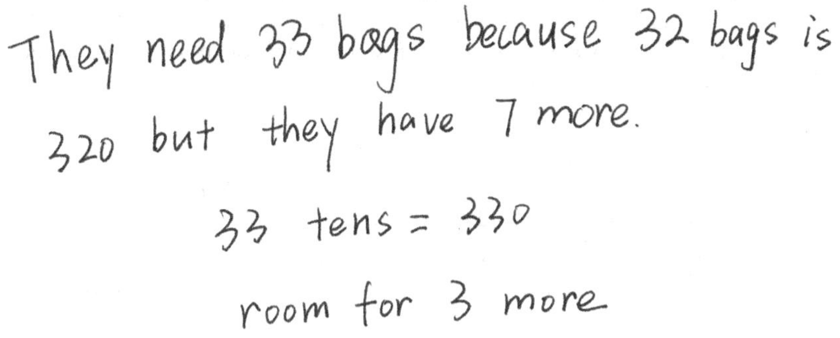

They need 33 bags because 32 bags is 320 but they have 7 more.

33 tens = 330

room for 3 more

The *OGAP Addition and Subtraction Progressions*

The *OGAP Additive Framework* includes progressions for both addition and subtraction. However, the two progressions are tightly connected with identical structure, sharing the same levels but differing in the strategies students might use to solve tasks. Note that students often use addition to solve subtraction problems, so both progressions can be used to examine student strategies for situations involving subtraction. While the focus of this section is on the addition strategies, you can look at the *OGAP Subtraction Progression* to see related subtraction strategies. The *OGAP Subtraction Progression* will be discussed in detail in Chapter 7.

The *OGAP Addition and Subtraction Progressions* are designed to be used by teachers to gather evidence of student thinking and strategies related to addition and subtraction as well as identify underlying issues and errors that may impede a student's ability to expand their thinking when solving addition and subtraction tasks. Information on specific *Underlying Issues and Errors* (shown at the bottom right) should be gathered in order to address these issues instructionally in building students' fluency with addition and subtraction.

The *OGAP Addition Progression* levels illustrate the continuum of strategies students use to solve addition problems as they develop more sophisticated understanding of number and operations. The levels range from *Counting* to *Transitional* to *Additive* strategies that are evident in student work and thinking as they develop their understanding and fluency with whole number addition.

Early Counting Strategies

When students first begin to make sense of situations involving addition, they often treat it as a counting exercise. The student may use drawings or concrete models to first count out each number and then count everything again to determine a total, thereby counting three times. Hazel's response in Figure 2.10 shows evidence of understanding the task by modeling each number in the problem with circles and then counting all of the circles one by one. Hazel indicates each addend by circling that many counters but then determines the total by counting them all, starting at the beginning of her sketch. This strategy is evidence of thinking at the *Early Counting* level because she has modeled each quantity and counted all to find the total. The task involves three addends, which may have influenced the fact that she needed to draw all the addends out. When tasks are more complex, students will often use strategies that are lower down on the progression.

Figure 2.10 *Early Counting* strategy. Hazel models and counts all to solve a single-digit addition task with three addends

Shawn started to build a building with 6 blocks. Armando added 3 blocks onto the building. Then Shawn put 4 more blocks on top. How many blocks did Shawn and Armando use for the building?

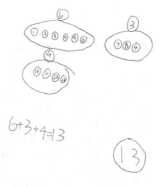

From Early Counting to Counting Strategies

As students develop stronger understanding of number, they employ strategies that indicate they can count on from a number and recognize the value of counting on from the largest number, using a variety of physical and visual models. This allows for more flexibility and a transition towards more efficient strategies.

In Figure 2.11 Darren uses a student-generated number line to count on to determine the total number of crayons. His jumps on the number line indicate that he counted by ones. He keeps track of the 9 he is adding on to 15 below the number line. Importantly, he does not need to model the quantity of 15, only the 9 that he is adding on. He does not label his answer, but he writes an equation that matches his work.

Figure 2.11 *Counting* strategy. Darren creates an accurate number line to help him count on from 15

Coral and Sean are counting crayons. Coral counted 15 blue crayons. Sean counted 9 red crayons. How many crayons did they count together?

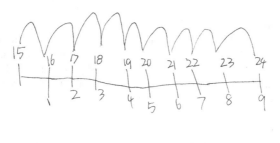

From Counting to Early Transitional Strategies

Base-ten understanding, the ability to compose and decompose numbers, and visual models that encourage unitizing can be used to move students from *Counting* to *Transitional* strategies. Students who are showing evidence of employing these ideas yet doing so inefficiently are often using *Early Transitional* strategies.

In Figure 2.12, Jack's work is evidence of an *Early Transitional* strategy. His work indicates an ability to decompose the two quantities in the problem into tens and ones. He draws a base-ten model to break both numbers down to tens and ones (vertical lines to represent the tens and small boxes to represent the ones) and then counts up the tens and ones separately. While this strategy leads to an accurate answer and shows important understanding of addition and place value, it is inefficient.

Figure 2.12 *Early Transitional* strategy. Jack models using base-ten blocks and adds by tens and ones to find the solution

Hasan played tag for 25 minutes and then played basketball for 28 minutes. How many minutes did Hasan play altogether?

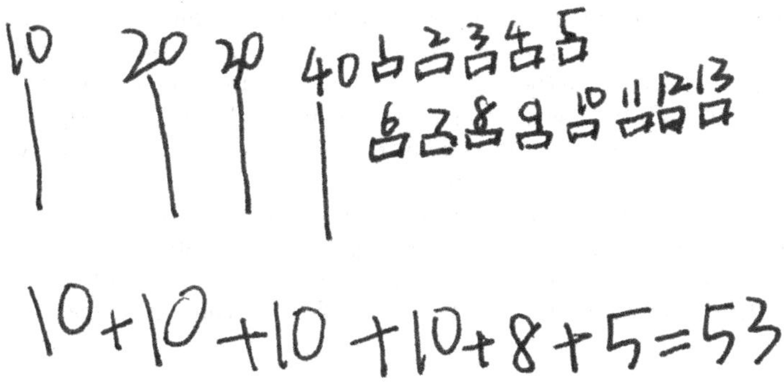

$$10 + 10 + 10 + 10 + 8 + 5 = 53$$

From Early Transitional to Transitional Strategies

The distinctive feature between *Early Transitional* and *Transitional* strategies is a shift to working with multiples of ten and employing more efficient strategies, while still relying on visual models. In Figure 2.13, instead of decomposing 53 into 5 tens, Lola demonstrates the ability to work with larger chunks of the 53. It is likely that she broke the 50 into 20 and 30 to help her add over 100.

Students whose work shows evidence of *Transitional* strategies have begun to think more flexibly and at the same time use base-ten understanding with models that support these efficient strategies. Empty number lines—that is, number lines that do not contain every number—are a particularly helpful model for students using *Transitional* strategies, as they offer the opportunity to work with numbers in multiples of tens or hundreds.

Figure 2.13 *Transitional* strategy. Lola adds on 53 using multiples of ten on a number line. She labels her answer indicating she understands the meaning of her answer

Hakim had a collection of 92 baseball cards. He got 53 new baseball cards for his birthday. How many baseball cards does Hakim have now?

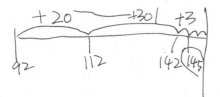

 Chapter 5 on Visual Models to Support Additive Reasoning to learn more about using empty number lines and base-ten block models for addition and subtraction.

From Transitional to Additive Strategies

Student work at the *Additive* level shows evidence of procedural fluency and flexibility based on the numbers and the situation. At this level, students' strategies have moved away from reliance on visual models towards efficient algorithms and strategies that have grown out of those visual models. In mathematics an algorithm is a procedure or set of efficient steps for solving a problem, such as the Partial Sums algorithm or the Traditional US algorithm for addition. As will be discussed further in Chapter 6, it is important that the use of algorithms be based on deep understanding of place value and the base-ten structure of the number system.

Student work at this level may reflect strong base-ten understanding by using algorithms based on place value understanding, as in Bruno's strategy in Figure 2.14. Bruno first adds the 100 from the 132 to the total, then adds 30 and 80 to get 110, and finally adds the 2 and 9 to get 11. Using efficient recording, he then adds the three partial sums to get the sum of 221. In Figure 2.15, Tasha's work shows evidence of a different *Additive* strategy, referred to on the progression as *flexible compensation*. Both students use strategies that make sense for the numbers and the situation.

Compare the two additive strategies in Figures 2.14 and 2.15 and consider how they are similar and how they are different. What could you say about these students' understanding of addition and base ten?

Figure 2.14 *Additive* strategy. Bruno uses a Partial Sums algorithm to solve the problem

132 students were watching the school play. Then 89 more students went to watch the school play. How many students are watching the school play in all?

$$\begin{array}{r} {}^{1}132 \\ +\;89 \\ \hline 100+110+11 \\ 221 \end{array}$$

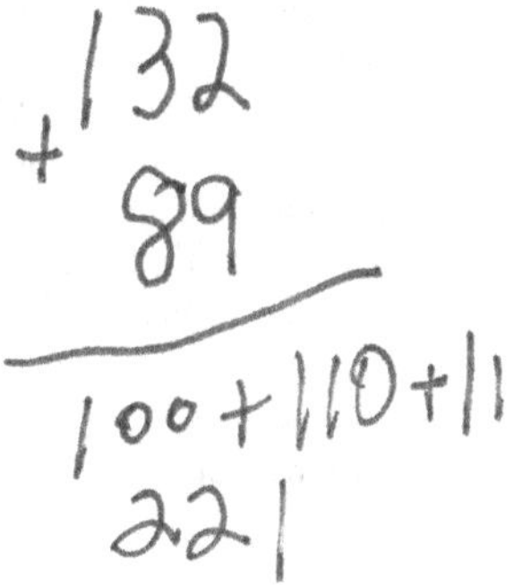

Figure 2.15 *Additive* strategy. Tasha uses a flexible compensation strategy to add 89 and 132

132 students were watching the school play. Then 89 more students went to watch the school play. How many students are watching the school play in all?

$$132 + 89 = 221$$
$$131 + 90 =$$
$$121 + 100 =$$

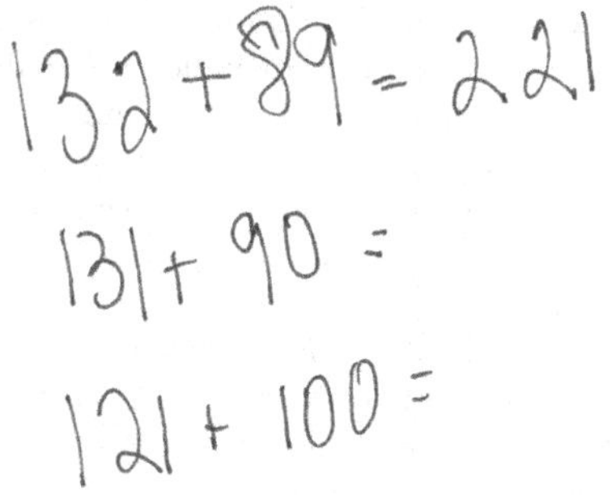

For more information on the Partial Sum algorithm, compensation, and other strategies and algorithms for addition see Chapter 6 Addition.

There is strong evidence in Bruno's use of the Partial Sums algorithm in Figure 2.14 that he can calculate accurately and has flexible base-ten understanding of the quantities in the problem. In Figure 2.15, Tasha's approach is also flexible because she is using compensation to make the problem easier to solve. She first takes one from the 132 and adds one to the 89, resulting in an equivalent expression of 131 + 90. She then takes 10 from the 131 and gives it to the 90 to make 100. The resulting expression, 121 + 100, is an easier problem to solve mentally and on paper.

While both of these students used *Additive* strategies to solve this problem, there is still room for further development of their additive reasoning. Research has shown that as tasks become more complex and the situations less familiar, student strategies will move up and down on the progression. As will be discussed further in Chapter 8 on problem solving, this means that it is important to vary the context of the problems, as well as the complexity of the numbers to further develop and strengthen student thinking and understanding. An important instructional step for both these students might be to choose a more challenging context or a task with larger numbers to extend and deepen their thinking.

 Chapter 8 Additive Situations and Problem Solving for more on this topic.

Non-Additive Strategies

Located at the bottom left of the *OGAP Addition Progression* is a level titled *Non-Additive Strategies*. Sometimes students use strategies that are not viable for the problem situation or will not result in a reasonable solution to the task. When they don't understand or access the problem situation, students may use an incorrect operation or model the problem incorrectly. Sometimes students attempt to use a strategy that they do not understand, and their mistakes illustrate that they have only a flawed procedural understanding (i.e., using a procedure without understanding). Occasionally there is a lack of evidence in the student work or the evidence seems to indicate that the student guessed at a strategy or an answer. Identifying a student solution as *Non-additive* is an indication that the student likely needs more instruction in order to access the task or develop a viable strategy. *Non-Additive* strategies reflect a variety of misunderstandings or issues, and students whose work is at this level may have very different instructional needs.

Examine student solutions to the same subtraction task in Figure 2.16. Both pieces of student work show evidence of *Non-Additive Strategies* on the progression. How would the focus of instruction be different for meeting the needs of these two students based on the evidence in their work?

Figure 2.16 *Non-Additive* strategies

Sia drove from Boston to New York. He left Boston drove 96 miles, and then stopped to eat. After lunch, he drove the rest of the way to New York. She drove 215 miles altogether. How far did Sia drive after lunch?

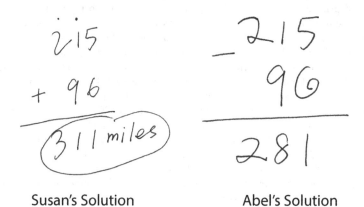

Susan's Solution   Abel's Solution

While addition can be used to solve this problem, and Susan's addition is correct, she uses the wrong operation for the numbers she is working with, showing evidence of using an incorrect operation for the task. In Figure 2.16 Abel sets up the correct problem to solve but uses the subtraction procedure incorrectly, resulting in an answer that is larger than the number he started with. These kinds of errors are often evident in student work when they have been introduced to a procedure or algorithm but do not yet have the underlying understanding of number or the operation to understand why the procedure works.

The Common Core State Standards for Mathematics (CCSSM)

The CCSSM represents a progression of developing procedural fluency related to number, addition, and subtraction using visual models, properties of operations, base-ten understanding, and the relationship between addition and subtraction. As detailed below, the CCSSM in grades K–2 focus on developing increasingly sophisticated strategies based on the use of concrete and visual models and deep understanding of number and base ten.

In kindergarten, students are expected to develop early number concepts through oral and written counting and using counting to solve addition and subtraction situations. Students also build understanding of number and number relationships by using concrete models, visual images, drawings, and verbal explanations and ultimately connecting many of these ideas to written number and equations.

In grade 1, students extend their understanding of number to include larger numbers and to begin to develop an understanding of the base-ten number system. They use their base-ten understanding to add and subtract larger numbers, such as adding and subtracting tens to a two-digit number. Grade 1 students are expected to use a variety of visual models and sketches to add and subtract, focusing on deepening their conceptual understanding of number and operations.

Grade 2 students are expected to extend their understanding of the base-ten system to include ideas of counting by tens, hundreds, and multiples of hundreds. They should understand multiple representations of multi-digit numbers and decompose numbers flexibly, depending on the needs of the situation and numbers being used. Students in grade 2 use their base-ten understanding and visual models to add and subtract multi-digit numbers, with an expectation that they can explain why addition and subtraction strategies work, using visual models, place value, and properties of operations.

Grade 3 students are expected to extend their conceptual understanding of addition and subtraction of multi-digit numbers to use strategies and algorithms for addition and subtraction that are based on place value, properties of operations, and the relationship between addition and subtraction. Notably, this is the first time any mention of an algorithm related to addition or subtraction is mentioned in the CCSSM. The CCSSM reflects a large body of research that suggests that premature teaching of abstract algorithms can be detrimental in the long run (Hiebert & Wearne, 1996; Kamii & Dominick, 1998). Rather, instruction in grades K–2 should focus on developing whole number addition and subtraction with understanding, as reflected on the progressions in the *OGAP Additive Framework*.

 Chapter 6 Developing Whole Number Addition and Chapter 7 Developing Whole Number Subtraction for more detailed discussion of algorithms.

The progressions contained in the *OGAP Additive Framework* reflect the importance and focus on developing a strong conceptual foundation for number and base-ten, addition, and subtraction. As in the CCSSM, all three progressions begin with anchoring number understanding on counting, move to developing strong visual models to make sense of concepts related to number, addition, and subtraction, and reflect the focus on strategies and algorithms based on place value and properties of operations. These ideas will be developed in more detail throughout the book.

Important Ideas about the *OGAP Additive Framework*

There are several important ideas to keep in mind when using the progressions that make up the *OGAP Additive Framework*.

1. **The *OGAP Addition, Subtraction,* and *Base Ten Number Progressions* reflect the mathematics education research on how students develop addition, subtraction, and base-ten number fluency with understanding.** They are meant to provide guidance to both teachers and students about possible next instructional steps in line with the expectations of the Common CCSSM (NGA & CCSSO, 2010).

2. **The three progressions contained in the *OGAP Additive Framework* are related to each other.** As students develop understanding of the concepts represented on the *OGAP Base Ten Number Progression*, they will also show increasingly sophisticated strategies on the *Addition* and *Subtraction Progressions*. Conversely, if a student is using *Early Counting* strategies to count and compare numbers, they will likely struggle to use *Transitional* or *Additive* strategies to add and subtract. The progressions for addition

and subtraction are also related because students will use inverse operations to solve problems (e.g., a missing addend problem can be solved by adding up from the given addend or by subtracting the addend from the total.) For this reason, when examining evidence in student work on subtraction tasks, looking at both progressions simultaneously is important.

3. **The progressions in the *OGAP Additive Framework* are not evaluative but are intended to be used for descriptive evidence and instructional decision-making.** Neither grade levels nor points have not been assigned to the various levels on the progressions; rather evidence in the student work should guide the instructional goals for each student.

4. **The progressions are designed at a grain size to be usable and manageable by teachers.** For this reason, not all possible strategies are represented on the progressions. If you find a strategy in student work that is not on the progression, you can look for structural commonalities in the strategies that are shown to determine the level of the student's thinking.

5. **The progressions provide instructional guidance.** The vertical arrow on the right side of each progression includes concepts and understandings that are critical for moving students to more efficient strategies along the progression. There is no one right path to move students towards more efficient and sophisticated strategies; rather the path is dependent on a student's strengths and weaknesses in understanding.

6. **Movement on the progressions is not linear.** The arrow that runs on the left side of each progression communicates the idea that as students encounter new concepts, interact with new problem situations and structures, or larger numbers, their strategies will move up and down along the progression. The ultimate goal is for students to develop a strong foundation for base-ten and additive reasoning by the end of second grade. This understanding will be built on in subsequent grades, as students are expected to demonstrate procedural fluency with addition and subtraction in upcoming grades.

7. **Paying attention to underlying issues and errors is also important.** At the bottom, on the right side of each progression is a box titled *Underlying Issues/Errors*. It is important to remember to keep track of errors in student work, as they often provide evidence of underlying issues that can impede student acquisition of concepts and strategies. For example, many of the student solutions shown in this chapter contain a numerical answer without a unit label. This can be an indication that the student does not understand the meaning of the quantities in the problem, or that further instruction on using labels in answers needs to take place. Classroom trends related to this information can give teachers actionable information about underlying concerns that may need to be addressed instructionally, either with the whole class, a small group of students, or an individual.

Using the *OGAP Progressions* as an Instructional Tool

OGAP is an intentional, content-focused formative assessment system, where learning progressions are used to analyze evidence in student work and inform mathematics instruction. Figure 2.17 illustrates the three key components of the *OGAP Formative Assessment Cycle*.

Figure 2.17 The *OGAP Formative Assessment Cycle*

As Figure 2.17 shows, formative assessment is an ongoing cyclical process. According to Black and Wiliam (2009), to be considered formative, the evidence must be "elicited, interpreted, and used" by teachers and learners regularly (p. 9). The first step is gathering evidence of student thinking around the learning goal. This can be done by administering an exit task at the end of a lesson. The next step is to analyze the evidence of student thinking reflected in the strategy used by using the appropriate learning progression. The final, and most important step, is to use the analysis of the evidence in the student work to make informed instructional decisions. Completing all three components of the cycle is critical for having the optimal effect on student learning and growth. After making instructional adjustments, the cycle begins again by selecting a method to gather more evidence of student understanding.

Giving entrance or exit questions at the beginning or end of a lesson are two examples of ways to regularly gather evidence of student thinking. Although teachers should be observing student thinking and adjusting responsively throughout the lesson, intentionally collecting written evidence of student thinking gives teachers an opportunity to reflect on student understanding without the immediate demand of an instructional response. Keeping the focus of the evidence collected on one or two important aspects of the learning goals will allow teachers to analyze the work quickly and formulate an instructional response that is timely. Ideally, the entire cycle, including an instructional response, should take place by the next instructional day. This allows for instructional adjustment throughout a unit of study, rather than waiting until a unit test or quiz is given, when it is often too late to respond productively.

Using the *OGAP Sort*

During the *Analyze Evidence* phase of the formative assessment cycle, teachers analyze the student work collected using the progressions through an *OGAP Sort*. This provides an opportunity to gather evidence to inform instruction related to three areas:

1. The level of the strategy used on the progression.
2. Underlying issues and errors in the solution.
3. The correctness or accuracy of solution.

The focus of an *OGAP Sort* is to first look at the level on the progression that best reflects the evidence in the student work. This is most easily done by sorting the work into separate piles that correspond to the levels on the progression, as shown in Figure 2.18. Figure 2.18 illustrates the general categories on all three progressions in the *OGAP Additive Framework*. A teacher may want to additionally sort into the sublevels within each of the categories (e.g., *Early Transitional* strategies and *Transitional* strategies). Using this level of detail when sorting student work can be helpful when planning for instruction.

Figure 2.18 Sorting student work into piles based on the *OGAP Progressions*

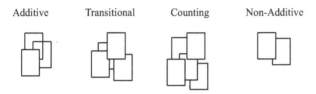

Once the work has been sorted into piles, the information can be recorded on the *OGAP Additive Evidence Collection Sheet*, as shown in Figure 2.19. Students' names are listed under the category that best represents the strategy they used to solve the problem as well and under the corresponding underlying issue or concern area if it was evident in their work. A circle around a name is an indication of an incorrect solution.

While it is possible to make a quick plan for instruction without a written record, the collection sheet offers additional opportunity to reflect on the trends and compare the trends to evidence from other exit questions. Using the *OGAP Additive Evidence Collection Sheet* to record evidence for multiple questions makes it easier to bring trends to light and respond instructionally. A blank copy of the collection sheet is included at the end of this chapter.

Next, going back through the work to look for underlying issues and concerns, and recording those on the collection sheet, will allow the teacher to see trends in errors within a set of work and across several exit questions. These can be indications of misunderstandings that might otherwise go unnoticed. For example, in Figure 2.19, the evidence collection sheet shows that many students did not include a unit label in their solution. This may be an indication that they do not know what their answer means in the context of the problem situation. The teacher may decide to address this as a whole group, a small group, or wait to see if the trend continues over several exit questions.

Finally, taking note of the correctness of the solution provides an additional piece of information to be considered. It is important to note that correctness is not the first information gathered because it is not as informative instructionally. Knowing that a student has obtained the wrong answer does not provide any guidance as to what is needed to move the student understanding forward. It is more typical for teachers to sort student work by levels of correctness, but looking beyond correctness to focus on student strategies often provides more actionable information and valuable details for determining next steps.

Figure 2.19 A sample of an *OGAP Additive Evidence Collection Sheet* that has been used to record evidence in student work after an *OGAP Sort*

Student Work Data Collection Sheet   Addition and Subtraction Recording Sheet

Item #	Content (e.g. context, type of number)	Counting by Ones		Transitional		Additive		Non-additive Reasoning
		Early	Counting On/Back/Up	Early	Transitional	Transparent Algorithms	Efficient Algorithms	
	Add to Result unknown Double Digit		Mya Paul Win	Ada Lin Eli Jack Kim	Greg Hakim	Derek Finn Ila	Ben	Cara

Underlying Issues or concerns								
Unreasonable	Place value error	Units inconsistent or absent	Property or relationship error	Calculation error	Equation error	Model error	Vocabulary error	
	Cara	Cara Finn Greg Ila Win				Eli Jack		

Once the evidence is recorded, the next step is to analyze the results and consider the best instructional response for the class. Asking the three questions below will help to complete the *OGAP Formative Assessment Cycle*.

1. What are developing understandings in the student work that can be built upon?
2. What issues or concerns are evidenced in the student work?
3. What are potential next instructional steps for the whole class, small groups, or for individuals?

The first question is focused on considering the good news that is evident in the student work: what students understood and were able to do. The evidence of student understanding can be used to anchor instruction and is often an indication of where a teacher's instructional response can begin. The second question asks teachers to consider what issues or concerns are evident in the work and often points to areas that the instructional response should focus on. These two questions together provide the information a teacher will use to design their instructional response.

After completing the cycle, a teacher would begin the cycle again by deciding on a new item to give and then collecting, analyzing, and responding to evidence in student work. The goal is to engage in this process regularly over time to continually improve student learning. Research by Supovitz and colleagues (2018) shows a positive and significant impact of the *OGAP Formative Assessment Cycle* on student achievement.

Most chapters in this book include multiple references to the *OGAP Additive Framework*. *The OGAP Base Ten Number, Addition,* and *Subtraction Progressions* are referred to throughout chapters when looking at student work and also when discussing strategies and algorithms used for number and operations. It will be helpful to have a copy of the *OGAP Additive Framework* available to use when any part of the framework is referenced.

Chapter Summary

This chapter focuses on the *OGAP Base Ten Number Progression* and the *OGAP Addition Progression*. As indicated previously, the progression for subtraction is closely related and will be discussed more explicitly in Chapter 7.

- The *OGAP Additive Framework* consists of four parts: *Problem Structures* (discussed in Chapter 8), the *Addition Progression, Subtraction Progression,* and *Base Ten Number Progression.*
- The progressions are based on the mathematics education research and are developed at a grain size that is usable by classroom teachers and students when working on concepts of number, base ten, addition, and subtraction.
- The progressions are designed to focus on strategies students use to solve problems and are meant to provide guidance about instructional steps that are targeted to students' individual needs and move all students forward mathematically. Sorting and analyzing work using the progressions helps to focus teaching and learning on the concepts that are foundational for the development of additive reasoning.
- OGAP is a content-focused, formative assessment system that involves implementing an ongoing cycle with three parts: Collect evidence, analyze evidence, and respond to evidence. In order to truly be formative in nature it is essential that all three parts of the *OGAP Cycle* be completed in a timely manner. Engaging in the cycle provides important information about where students are in their understanding and what the next best instructional response may be.
- The progressions are designed to illustrate how student strategies develop and progress from *Counting* (by ones at first and later unitizing in groups) to *Transitional* (with larger groups and visual models) to more abstract and efficient *Additive* or *Base Ten* strategies. As students are exposed to more complex numbers and situations their strategies will move up and down the progressions.

Looking Back

1. **Become Familiar with the *OGAP Additive Framework*:** The framework is composed of four sections, *Problem Structures* on the front page, the *Base Ten Number Progression* on the back page, and the *Addition Progression* and *Subtraction Progression* on pages 2 and 3.
   (a) Review the three progressions found in the framework. How does the grain size of the information presented make it useful for classroom teachers when making instructional decisions?
   (b) When looking at the *OGAP Addition and Subtraction Progressions* on the inside of the framework, what connections exist between the levels on both progressions? Why do these similarities exist?

    (c) What is common to strategies at the *Transitional* levels on all three progressions?

2. **Collecting Evidence of Student Understanding:** As an approach to formative assessment, *OGAP* is cyclical in nature and consists of three components. Gathering evidence, the first component, can be done in a variety of ways including entrance tickets or exit tickets.

    (a) Why is it important to gather written evidence of individual student understanding?

    (b) What is the benefit of regularly collecting evidence that is focused on only one or two important aspects of a lesson?

Figure 2.20 The *OGAP Formative Assessment Cycle*

    (c) Information about a student's developing understanding can be useful in a variety of school situations. Beyond instructional decision making, in what other ways can the evidence be used?

3. **Using Student Work to Inform Instruction:** Imagine a teacher uses the following problem as an exit ticket.

Write the number that is 20 tens and 9 ones. Explain how you know.

    (a) What makes this task useful for assessing students' understanding of base ten? What are some things you would want to look for in students' responses?

    (b) Use the *OGAP Base Ten Number Progression* to analyze the following student responses to this task. For each piece of student work identify the level on the progression the work best reflects and any underlying issues or errors.

    (c) Use this analysis to formulate an instructional response for each student based on the evidence.

Figure 2.21 Four Student Responses

Write the number that is 20 tens and 9 ones. Explain how you know.

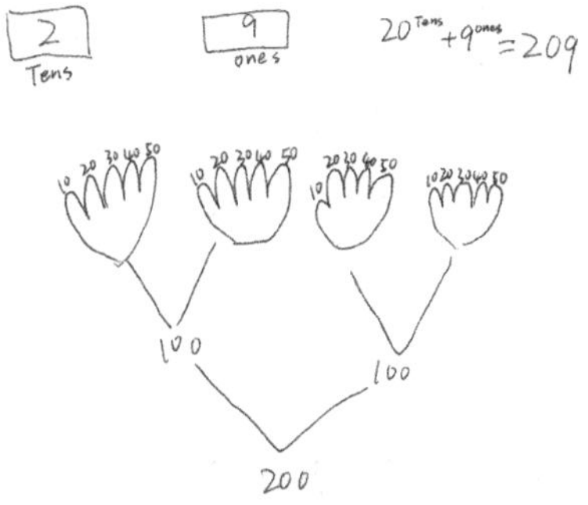

$$\boxed{2}\ \text{Tens} \qquad \boxed{9}\ \text{ones} \qquad 20^{Tens}+9^{ones}=209$$

200

Student A

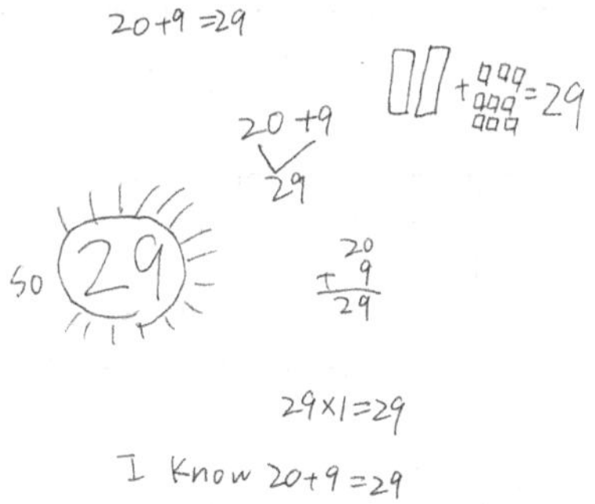

$20+9=29$

$20+9$

29

$50 \quad 29$

$\boxed{}\boxed{}+\begin{smallmatrix}9\,9\,9\\9\,9\,9\\9\,9\,9\end{smallmatrix}=29$

$\begin{array}{r}20\\9\\+\\\hline 29\end{array}$

$29\times1=29$

I know $20+9=29$

Student B

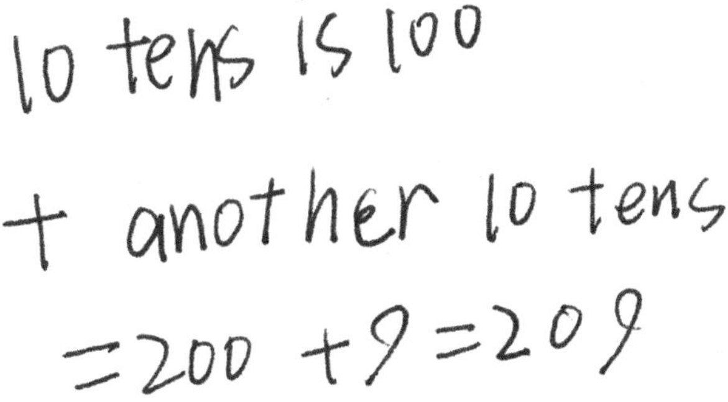

lo tens is 100

+ another lo tens

$=200+9=209$

Student C

Figure 2.21 Continued.

Student D

Instructional Link

Use the following questions to analyze ways your instruction and mathematics program provide regular opportunities to collect descriptive formative evidence of student understanding.

1. Are there resources or processes for monitoring the development of student understanding over time and over a variety of problem structures, situations, and magnitude of number?
2. Based on this analysis, what are some specific ways you can enhance your mathematics instruction using ideas from this chapter?

Figure 2.22 *OGAP Evidence Collection Sheet*

Student Work Data Collection Sheet — Addition and Subtraction Recording Sheet

Item #	Content (e.g., context, type of number)	Counting by Ones		Transitional		Additive		Non-additive Reasoning
		Early	Counting On/Back/Up	Early	Transitional	Transparent Algorithms	Efficient Algorithms	

Underlying Issues or concerns

Unreasonable	Place value error	Units inconsistent or absent	Property or relationship error	Calculation error	Equation error	Model error	Vocabulary error

3

The Development of Counting and Early Number Concepts

Big Ideas

- For young children, learning to count often begins with rote memory of the counting sequence. Over time children develop the ability to count a set and then to use counting to solve problems.
- Strategies for using counting to solve problems become more sophisticated as children develop understanding of foundational concepts. Over time and with support, students will transition from counting all, to counting on, to counting in groups, and finally to utilizing their understanding of the base-ten number system to count and compare.
- Procedural and conceptual competence in counting and comparing quantities develops over time through engaging in a variety of purposeful counting activities.
- By observing students counting and solving problems, teachers can gather evidence about what they already know about number and counting and use this to inform instruction.

The Development of Number

For young children, learning to count involves both knowing the counting sequence and understanding foundational concepts. As discussed in Chapter 2, the *OGAP Base Ten Number Progression* illustrates how children's developing understanding of number is reflected in the strategies they use to determine quantity and locate and compare quantities. Before they can count with meaning, children often learn the counting sequence by rote, as a memorized string of words. Early counting involves recognizing patterns in this sequence and being able to use this sequence to count a set of objects one by one and compare and locate numbers on a number path or number line. Transitional strategies involve counting in groups, initially with visual models where ones are visible, and using groups to compare and locate numbers. This builds an important bridge to being able to use place value and base-ten understanding to compose and decompose, locate, and compare numbers of any size. The development of base-ten number understanding is critical for supporting the development of flexible addition and subtraction strategies (see Chapter 6 and

Chapter 7) as well as multiplication and division (see *A Focus on Multiplication and Division*). This chapter explores the foundational concepts that support this development in the early grades. Chapter 4 follows by focusing specifically on the development of base-ten number understanding. Early number concepts develop and deepen through experiences with counting to solve problems in a range of contexts and situations. It is important to remember that these concepts (1) are developed first with smaller quantities; (2) do not necessarily develop in a particular order or at set ages; and (3) are often related and develop concurrently.

Early Counting: Learning to Count and Compare Sets

Very young children can determine how many objects are in a set of two or three objects without counting. This perceptual ability, called *subitizing*, is an important foundation for developing understanding of number and is discussed later in the chapter. In order to determine the numerosity of a set of more than four objects, children need to learn to use counting to determine the total amount.

Counting a set of objects to determine the total amount involves coordinating three logical principles: reciting the counting sequence in the correct order, attaching one number word to each object, and recognizing that the last number said names the quantity of the set (Gelman & Gallistel, 1978). Each of these principles is discussed separately, though they often develop concurrently.

Knowing and Understanding the Counting Sequence

The counting sequence is a social convention, as different cultures use a variety of counting systems and symbols. We use a base-ten counting system; the numerals 0–9 are used to represent quantities as groups of tens and ones, and then groups of hundreds, tens, and ones, and so on. The spoken words in the English language make it hard to see this grouping and regrouping in numbers below 100, especially in the decades and teen numbers, and most children need to memorize the number words up to 12. Although the number words from 13–19 and the decades provide some clues to the base-ten structure, the connection is not always straightforward (e.g., it is not obvious that fifteen is "five and ten" or thirty is "three tens"). In addition, the order of putting the word for ones before the word for tens is opposite to the way the numerals are written (sixteen vs. 16). In other languages the connection to the base-ten place value structure is much more explicit. For example, in Japanese, fifteen is *jugo* which is literally "ten-five" and thirty is *sanju* which is literally "three-ten."

Very young children often learn to recite the counting sequence verbally from memory as if it is a list of words or a song, called *rote counting*. Over time, and as they gain experience with numbers over 20, they begin to recognize patterns in the structure of the number system, such as the idea that the number words from 0 to 9 repeat in a predictable pattern. However, as they start to notice patterns, children sometimes make errors (e.g., saying. "eighty-ten" instead of "ninety"). Importantly, these errors are actually signs that students are starting to make sense of the base-ten structure of the counting sequence.

One-to-One Correspondence

When children learn to attach one number word to one object while counting, they have developed the principle of *one-to-one correspondence*. While children are still developing this principle, you might see them double counting one object, or reciting the counting sequence without matching each spoken number word to a single object. These strategies are at the pre-counting level on the *OGAP Base Ten Number Progression*. One-to-one correspondence does not develop all at once or in the course of one activity or lesson, and often you will see evidence of partial or incomplete understanding in children's counting activity (Carpenter, Franke, Johnson, Turrou, & Wager, 2017).

Young children are initially more successful with using one-to-one correspondence to count objects when they are presented in a line (Fuson, 1988). In order to count objects in a random arrangement, they need to use one-to-one correspondence and also need to keep track of the objects they have counted. With physical objects, they can learn to use the strategy of touching and moving an object to the side after counting to keep track (tag and count). When counting pictures rather than objects, they have to devise a system to keep track, for example, by writing tick marks or numerals next to each picture. In Figure 3.1, you can see how each child uses a different strategy to keep track while counting 18 blocks represented on paper, in a random arrangement. Aidan circles each block and labels it with a numeral, showing solid understanding of one-to-one correspondence. Brian uses check marks to keep track while counting verbally, and Isaiah draws a path through the blocks. While Isaiah recognizes the need to keep track, the evidence suggests that he has made a mistake in the number sequence and also double counted one object.

Figure 3.1 Three student strategies for counting a collection of objects

How many blocks are there?

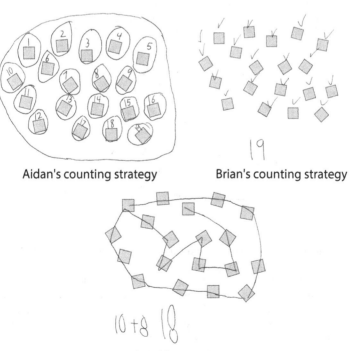

Aidan's counting strategy   Brian's counting strategy

Isaiah's counting strategy

Cardinality

The third important counting principle is *cardinality*, an understanding that when counting, the last number word said refers to how many things there are in the whole set. When young children are asked the question "how many" after counting a set, they will sometimes go back and recount, indicating that they consider the answer to be the counting sequence itself: "1, 2, 3, 4, 5, 6" rather than the last number said (6). This is evidence that the student understands counting as an activity, but has not yet constructed cardinality, an understanding that counting can determine quantity. Some students may seem to understand that the last word said answers the question "how many" without grasping the more abstract idea of cardinality. For example, as illustrated in Figure 3.2, they may say "6" to answer how many pencils are in the set but when asked to show the 6 pencils they point to the last pencil, indicating that they think 6 is the name for the last pencil counted rather than the amount in the entire set (National Research Council, 2009).

 After a child has counted a set of objects, ask "how many are there?" If the child repeats the counting sequence or answers with a number word different than the last number counted, this may be evidence that they have not yet connected counting and cardinality. If they do answer 6, ask them to show you the 6 to confirm their understanding of cardinality.

Figure 3.2 A child with cardinality understands that 6 refers to the whole quantity of pencils, rather than being a name for the sixth pencil

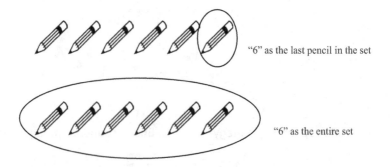

Once children have developed cardinality, a slightly more complex task is being able to produce sets of a given quantity. Producing a set is more difficult because it requires working in the opposite direction—remembering the given quantity mentally while counting out to that number. In Figure 3.3, Gabriela and Katie have been asked to produce and draw a tower of 8 cubes. Gabriela draws cubes, but is unable to show that she has 8. Katie draws a tower of 8 cubes and shows *one to one correspondence* in her counting to show that there are 8.

Figure 3.3 Producing and drawing 8 cubes

Build a tower with 8 cubes. Draw a picture of your tower. How do you know your tower has 8 cubes?

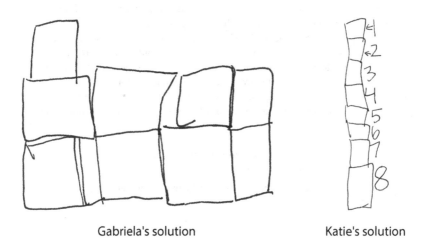

Gabriela's solution   Katie's solution

These three counting principles (the counting sequence, one-to-one correspondence, and cardinality) are not developed in a predictable sequence but instead develop concurrently, and often at different rates, through experience in a range of contexts. For example, if a child has developed one-to-one correspondence but does not have command of the number sequence past a certain number, their ability to count will be limited. In Figure 3.4, Chloe's work shows evidence of one-to-one correspondence and knowledge of the counting sequence up to 10, but she does complete the count to produce an answer to represent the total amount of blocks. It is likely that she only knows the count sequence up to 10 and this limits her ability to determine the cardinality of the set.

Figure 3.4 Chloe's solution shows evidence that she may not know the counting sequence beyond 10

How many blocks are there?

Likewise, if a child has memorized the counting sequence but does not have one-to-one correspondence, their counts will be inaccurate. Observing children as they count is critical for determining which principles are still developing and the next appropriate instructional steps. These three concepts: knowledge of the counting sequence, one-to-one correspondence, and cardinality need to be intentionally extended and developed in tandem for increasing quantities.

The examples shown in Figure 3.5 illustrate how children need to coordinate these three principles in the activity of counting. Ethan shows understanding of the counting sequence but fails to preserve one-to-one correspondence, while Hailey shows one-to-one correspondence but does not know the counting sequence from 15 to 18. Hailey also shows evidence of understanding cardinality in writing the answer of 19 to answer the question. As you can see, both Ethan and Hailey are still learning how to write numbers.

Figure 3.5 Ethan and Hailey's solutions to a counting problem show different understanding of the counting principles

Omar built 2 towers. How many blocks did he use altogether?

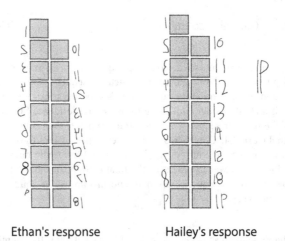

Ethan's response   Hailey's response

Initially, the ability to correctly count a set of objects is dependent on the size of the set. Once their understanding of the counting sequence grows, and they have developed strong one-to-one correspondence, children can successfully count large groups of objects in any arrangement. Research shows that experience with counting larger sets is also important: sometimes children will show evidence of one-to-one correspondence when asked to count larger collections but not with smaller ones.

A child who is asked to give every student in the room a snack, and counts the number of children in the room first to determine how many snacks are needed before passing them out, is integrating the understanding of the counting sequence, one-to-one correspondence, and cardinality. There are many real-life activities and routines in the classroom that can support the use of these ideas, such as passing out materials, taking attendance, collecting data, or taking an inventory of classroom supplies. For more examples of classroom routines that incorporate early number concepts, see Fosnot and Dolk (2001), Franke et al. (2018), and Shumway (2011).

Once children have developed solid understanding of these three principles, the next step is to learn to use counting to solve problems, shifting from knowing *how* to count to knowing *when* to count (Nunes & Bryant, 1996). Initially children will solve problems by treating the quantities in the problem as collections of ones, as shown in the *Early Counting* level on the *OGAP Base Ten Number Progression*.

Counting by Ones to Solve Problems: Developing Flexibility and Efficiency

As children develop understanding of the counting sequence, one-to-one correspondence, and cardinality, along with the ability to record numbers in written form, they have the skills and strategies they need to use counting to solve a variety of problems (determining quantity, comparing quantities, locating numbers on a number line, and addition and subtraction) with increasingly larger quantities. Their problem solving will become more efficient and flexible as they come to the understanding that a number can represent the result of a count (or quantity), without actually having to do the counting. When young children are asked to solve a simple addition problem such as, "If you have 4 candies and I give you 3 more, how many candies will you have?" they will use counters or their fingers to *count all*. First they will count out four, then count out three more, and then count the whole collection again starting at one to find the total. Over time, they realize that they can begin with four and *count on* three more (5, 6, 7), a much more efficient strategy (Carpenter & Moser, 1984). More efficient strategies like counting on or back from a number to solve problems involve understanding of several additional foundational concepts: *hierarchical inclusion, conservation of number*, and *conceptual subitizing*.

Because these concepts require what Piaget called *logico-mathematical* knowledge, they cannot simply be shown or explained to students (Kamii, 1985). Children may be able to repeat or paraphrase these ideas back if the teacher explains them, but to truly understand these concepts, they must construct understanding of the relationships on their own. This requires mental actions such as reflecting on patterns and relationships, generalizing, and developing understanding of why the pattern happens.

Hierarchical Inclusion

Hierarchical inclusion is an understanding that a quantity grows by one with each count and therefore that the value of a number contains all the previous values in the counting sequence, as shown in Figure 3.6. If one more is added to four to get five, then one removed from five would make four, or nested inside the five is a four (and also three, two, and one) The ability to think about how 5 can be separated into 4 and 1 as different parts but also put back together make whole of 5 requires logical inference (Fosnot & Dolk, 2001). This understanding is important for developing the strategies of counting on or back from a given number and for addition and subtraction.

Figure 3.6 An illustration of the nesting of the values of numbers, or hierarchical inclusion

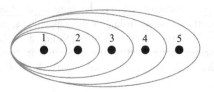

 To determine if a child has developed the concept of hierarchical inclusion, after they have counted a set, ask, "What if I took one away, do you know how many you would have then?" If the child can answer correctly without recounting the set, this is an indication that they have constructed hierarchical inclusion.

Mica shows evidence of this understanding, along with an understanding of the operation of addition, in her explanation shown in Figure 3.7. She understands that 6 + 7 must be greater than 6 + 4 and justifies this by showing that 7 is three units or jumps more than 4 on the number line. Her work also shows how important hierarchical inclusion is for understanding a counting on strategy for addition.

Figure 3.7 Mica's solution shows understanding of hierarchical inclusion

Bailey knows without adding that 6 + 7 is greater than 6 + 4. How does she know?

The big idea of hierarchical inclusion is extended into a more complete understanding of number and the construction of two more big ideas discussed later in this chapter: compensation and part–whole understanding. However, in order to develop these concepts, children must first have an understanding of *conservation*, the idea that a quantity is stable if nothing is added or taken away.

Conservation of Number

Conservation refers to the ability to determine that a certain quantity will remain the same despite adjustment of the container, shape, arrangement, or apparent size. Children are said to conserve number if they are aware that when two sets have been shown to be equivalent, either by one-to-one correspondence or by counting, this equivalence is maintained even when one of the sets is rearranged. Conservation involves recognizing that the number of objects is independent of the arrangement and the order in which the items are counted (i.e., beginning the counting from any item will produce the same result).

When asked to compare two sets of objects arranged in rows, young children may consider their visual properties rather than count both rows to compare the quantities. Consequently, when one row is compacted or pushed together, children without conservation of number will say that the longer row contains more objects, even when asked to count the objects in both rows. At this stage of development, they are relying more on their perceptions than on logical operations or understanding of quantity (Piaget, 1965). Another way to determine if a child has developed conservation of number is to rearrange a set of objects after they have counted them and determined the quantity. If the child has to recount the collection after the items have been rearranged in plain view without removing any of them, it is a sign that they lack conservation of number.

 To determine if a child has conservation of number, show them two rows of objects and ask, "Which row has more, or do they have the same amount?" Then spread one row out and ask the same question (see Figure 3.8). A child who is not yet conserving number will answer that the second row has more, sometimes even when asked to count both rows.

Figure 3.8 A conservation of number task: Which row has more, or do they have the same amount?

Conservation of number is one aspect of the larger concept of conservation, a developmental accomplishment in which the child understands that changing the form of a substance or object does not change its amount, overall volume, or mass. Conserving number is an indication that children can use logical thought or operations while they interact with physical objects (Piaget, 1965). The understanding that the cardinality of a set of objects is stable as long as no objects have been added or taken away allows children to develop more sophisticated strategies for adding and subtracting numbers without having to recount or model every object.

Once children have developed cardinality, hierarchical inclusion, and conservation, they are able to understand the strategy of *counting on from one number* rather than always starting from one. In order to count on, children need to be able to think about the first quantity abstractly (as a cardinal quantity), be able to generate the counting sequence from that number, and keep track of how many they have counted. As Fosnot and Dolk (2001) state, "they almost have to negate their earlier strategy of counting from the beginning" (p. 36). The student who solves 5 + 3 by counting all—counting out five, counting out three and then going back to count the total starting from one—is unable to hold the total 5 while they work on adding 3. In essence, the first quantity disappears once they begin to think about the second one. Students who count on their fingers and do not have a counting on strategy often struggle to solve problems greater than 10 because they cannot model all three quantities.

 To determine if a student has developed understanding of counting on, ask the student to count a set and then cover it. Then place additional objects next to it and ask how many there are in all. Does the student go back to try to count the first set again, or can they build from the cardinality of the previously counted set without actually seeing it?

In the examples of student work shown in Figure 3.9, Joy needs to draw out circles to represent the 6 crayons that are inside the bucket and the 3 outside the bucket and count them all out to find the total of 9. David is able to think about and count on from the quantity of 6, even though it is not visible in the bucket, to determine the total number of crayons.

Figure 3.9 Two student strategies for finding out the total when one quantity is hidden

There are 6 crayons in the bucket and 3 more on the table. How many crayons are there altogether? Show how you know.

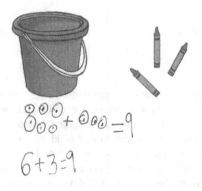

Joy's strategy

Figure 3.9 Continued.

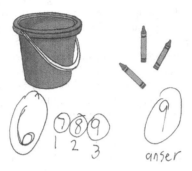

David's strategy

Generally, students learn to count on from the first number and then transition to counting on from the larger number. When making this shift they are applying their understanding of relative magnitude (discussed later in this chapter) and the *commutative property* (3 + 5 = 5 + 3) in order to use a more efficient counting strategy. The commutative property of addition is discussed in more detail in Chapters 1, 6, and 9.

Counting on or back from any number in the number sequence is an important skill that involves both conceptual understanding and procedural fluency. In general, counting back is more difficult for students, and therefore important to practice. In Figure 3.10, Jean shows understanding of the patterns in the number sequence but struggles to cross over the decade from 31 to 29 and then again from 21 to 19 when counting backward. Fluency when counting backwards is important for developing flexible subtraction strategies.

 Go To Chapter 7 Developing Whole Number Subtraction for more about flexible subtraction strategies.

Figure 3.10 Jean counts backward from 35 to 15 and makes errors when crossing the decade

Count backwards starting at 35. Stop at 15.

35, 34, 33, 32, 31, 20, 2 9, 28, 27, 26, 25, 24, 23, 22, 21, 19, 18, 17, 16, 15

Subitizing

Another skill that supports counting on is *subitizing*. Subitizing is the ability to recognize the numerosity of a group quickly and visually and connect the quantity to
the number name. Even very young children can look at very small collections or
groups and almost instantly tell how many there are without having to count them.
Research shows that the ability to recognize two objects as being different from
a single object is present even in infants, and most children can subitize up to four
objects by the time they enter school. This ability can be supported and extended
by providing students with experiences where they focus on quickly determining an
amount without counting, an activity that is often called Quick Images.

 Show or flash a dot image for 2 to 3 seconds and ask children to tell you
how many dots they see. Start with familiar patterns with 3 or 4 dots and
gradually add more dots and less familiar patterns or combinations.
Watch to see if children are moving their eyes or fingers to try to count
by ones.

When children are able to "just see" how many objects there are in a small set
then they are using *perceptual subitizing* to determine the quantity without counting. Most people are not able to perceptually subitize more than five objects unless
they are in a recognizable arrangement (e.g., six as two vertical rows of three on
a die). However, we can determine the total of a larger set—beyond the limits of
perceptual subitizing—by breaking the group into smaller subgroups. For example,
in Figure 3.11 the child can use the ten frame to see that there is a group of five
and a group of three and very quickly determine that there are eight dots. This is
called *conceptual subitizing*.

Figure 3.11 Conceptual subitizing using a ten frame

Conceptual subitizing helps to move children away from a reliance on counting
physical models toward internal or visual representations and mental strategies.
Conceptual subitizing is an important foundation for adding and subtracting small
numbers and supports more advanced strategies for computation. Therefore

children who do not have enough experiences to support the development of conceptual subitizing are at a disadvantage when learning many number and arithmetic processes.

Conceptual subitizing can be supported by regularly presenting children with opportunities to determine the total in sets arranged in patterns that can be easily recognized, such as the dot patterns found on dice. For example, in the pattern shown in Figure 3.12, a student might say, "I know there are 8 because I saw 4 and another 4." This is the foundation for understanding that 4 + 4 = 8 or 2 × 4 = 8. Another student might see the groups of two and count: 2, 4, 6, 8.

Figure 3.12 An image with familiar visual patterns can be flashed on a projection screen or shown quickly to elicit different conceptual subitizing strategies

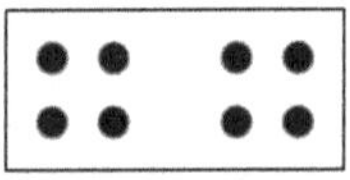

Sharing strategies for determining how many can help children see that numbers are composed of parts and wholes and develop understanding of both the additive and multiplicative composition of numbers. It is important to present or flash these images to children quickly (two or three seconds), in order to discourage counting by ones and encourage conceptual subitizing.

Ten-frames, structured bead strings or racks, and base-ten models can also be used to support conceptual subitizing.

Go To Chapter 5 Visual Models to Support Additive Reasoning for more on these models and others.

Young children can also use their fingers or *math hands* to develop conceptual subitizing with the numbers five and ten. In this activity, the teacher flashes an amount by holding up fingers and then ask students to name "how many," or the teacher can state a given number or task and ask students to use their fingers or math hands to show the quantity. This activity can be used to support many of the concepts discussed in this chapter. For example, the teacher can ask:

- What number is this? How do you know? (Begin with numbers less than 5, move to 5 and some more.)
- Show me 3. (To support both one-to-one correspondence and conservation, ask students to hold up their fingers next to another student to prove they both have 3.)
- Show me 5 using two hands. Find someone who is showing 5 in a different way. (Record the ways on the board to support the part–whole concept.)

- Show me two more than 5. (Watch to see if students start with 5 and then count up or just know it.)
- Show me one less than 7. (Watch to see if students know it is 6 without having to start at 7, demonstrating hierarchical inclusion.)

This activity can be extended to numbers greater than ten and to support base-ten understanding by teaching students to flash both hands to represent ten. More complex ideas such as multiplication and fraction concepts can also be supported with subitizing activities. (See *A Focus on Multiplication and Division* and *A Focus on Fractions* for more on subitizing in later grades.)

In sum, the concepts of hierarchical inclusion, conservation, and conceptual sub-itizing help students develop more sophisticated counting, addition, and subtraction strategies and are strengthened through meaningful activities that involve determin-ing and comparing quantities.

Comparing Quantities

Children develop the idea of magnitude before they construct cardinality (Fosnot & Dolk, 2001). The ability to subitize allows them to think about the size of a set and compare two sets perceptually before they are able to think about them in terms of quantity. If the sets are very different in size, they can rely on their perception to identify the bigger set. At this stage, they might also conclude that a set that cannot be perceptually subitized is bigger if it takes up more space. This is the beginning of an understanding of the more than/less than relations and is considered to be a *Precounting* strategy on the *OGAP Base Ten Number Progression*.

Once children have one-to-one correspondence, they can see that two sets have the same quantity by matching each item in the set to an item in the other set. If one set has left over items, then that set is greater. Children first learn to do this with concrete objects and then with pictures or drawings, as shown in Figure 3.13.

The task shown in Figure 3.13 was engineered to elicit student thinking around comparison. The three student responses show different strategies for determining who had more shapes. Seth answered that Will had more shapes. When his teacher asked him how he knew, he responded "it is longest" indicating that he was paying attention to spatial attributes rather than counting the objects. Rashmi uses match-ing to determine which set is larger; since there is one unmatched circle in Kai's set, he knows that set has more. Malik counts the objects to determine that Kai has more. His solution shows evidence of one-to-one correspondence and cardinality as well as knowledge of the counting sequence.

Figure 3.13 Three strategies for comparing quantities

Who has more shapes?

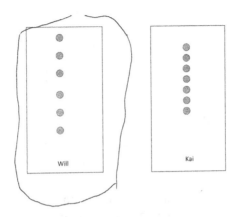

Seth's response, "It is longest."

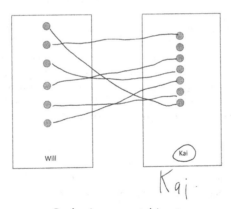

Rashmi uses matching to
determine who has more.

Figure 3.13 Continued.

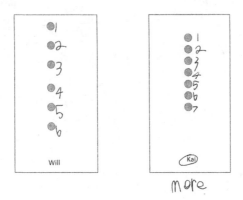

Malik uses counting to
compare the amounts.

Using counting to determine which set is greater by comparing the cardinal amounts is a significant development that grows out of understanding of the counting principles and hierarchical inclusion. Eight is bigger than seven because eight comes after seven in the counting sequence and eight is one more than seven.

Malik's use of counting to compare numbers by their cardinalities in Figure 3.13. is a strategy at the *Early Counting* level on the *OGAP Base Ten Number Progression*. At the next level, children are able to use counting to determine which set is more and also the difference between the smaller and larger set. Recognizing this third set as part of the larger set that tells how many more is important for being able to solve comparison problems in context. (See Chapter 8 Additive Situations and Problem Solving for more on problems that involve comparison.)

CCSSM on Counting

In the CCSSM, children are expected to develop the counting principles and number concepts discussed above by the end of kindergarten and extend this understanding to larger numbers in first and second grade.

- In kindergarten, children should learn the number names and the count sequence, count to determine the number of objects, and compare numbers. More specifically this includes counting to 100 by ones and tens, counting forward from a given number, producing a set of objects up to 20, and using counting and cardinality to answer how many are in a collection up to 20 objects in different arrangements (including a line, rectangular array, circle) and up to 10 things in a scattered arrangement. They are also expected to use matching and counting strategies to compare sets and compare written numerals between 1 and 10.

- In first grade, children extend this understanding to count, read, and write numbers to 120, and in second grade within 1000. In second grade, they are also expected to be able to skip count by 5's, 10's, and 100's and mentally add 10 or 100 to any given two- or three-digit number.

Composing and Decomposing Number by Parts and Equal Groups

As children develop more sophisticated understanding of number, they are able to see, compare, and operate on quantities not only as collections of ones, but as collections made up of parts and equal groups. This is a significant development towards additive reasoning and helps to build a foundation for base-ten understanding and calculating efficiently with larger numbers. The additional concepts that are important for developing this understanding include part–whole relationships and unitizing.

Part–Whole Relationships

When students understand hierarchical inclusion, they can see the number as a unit while at the same time seeing it as being made up of parts that reside inside the unit (e.g., 7 is made up of 6 and 1 more). Closely related to this idea is the *part-whole* concept, the understanding that any given number can be composed and decomposed into many different parts without changing the quantity of the whole. For example, 7 can be decomposed into 6 and 1, 5 and 2, or 4 and 3. This understanding requires being able to see that the parts reside inside the quantity as a whole and the reverse: a whole can be decomposed into its parts. Developing this concept, sometimes called *additive composition*, is essential for understanding the base-ten structure of the number system (e.g., understanding that teen numbers are made up of one ten and some ones) and for developing strategies for addition and subtraction. For example, knowing that 5 + 2 = 7 leads to the understanding that 7–2 is 5.

 Go To Chapter 1 Additive Reasoning and Number Sense for more on part-whole relationships.

Unitizing

Unitizing is the ability to see a collection of ones as both a group and a set of individual ones. This is a significant shift in thinking for children and is central to developing place value understanding—ten can be represented and thought of as one group of ten or ten individual units. Unitizing requires that children use the number sequence to count not only individual objects but also groups of objects—and to count them both simultaneously (e.g., one ten, two tens, three tens . . .).

 Go To Chapter 4 Unitizing, Number Composition and Base Ten Understanding for more on unitizing.

As Fosnot and Dolk (2001) explain:

> To construct an understanding of unitizing, children almost have to negate their earlier idea of number. They have just learned that one object needs one word — that one means one object, that ten means ten objects. Now ten objects are one — one ten. How can something be simultaneously one and ten?
>
> (p. 64)

Understanding of unitizing is a big idea or critical building block for the development of number sense. Children build the foundation for unitizing by counting things in small groups, like twos or fives, and developing the ability to skip count by these numbers with understanding. Counting by twos with understanding means that they not only know the skip count sequence from memory and pattern recognition (2, 4, 6, 8, 10, 12, 14 . . .) but understand that when they say "two" they are naming the quantity of a set of two objects. In the example shown in Figure 3.14, Gigi shows the ability to count the blocks by both ones and twos, using multiples of two for each group of two. This student understands that the blocks can be counted by ones or by groups of two.

Figure 3.14 Gigi counts the blocks by ones and by twos

Omar built 2 towers. How many blocks did he use altogether?

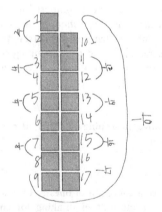

Knowing skip count sequences is an important skill, but children can memorize skip count sequences without having this underlying understanding of using skip counting to determine quantity. Unitizing is the ability to extend this understanding of counting by groups to understand how numbers can be composed of ones or units that are larger than one. In the example in Figure 3.14, Gigi shows the understanding that 17 can be composed of 17 ones or by groups of 2 and 1 more. As will be discussed in Chapter 4, unitizing by tens, and then later by 100, 1000, etc., is critical for understanding place value and the base-ten number system. For example, the number 35 can be thought of as 35 ones, or 3 units of ten and 5 ones.

Counting sets of objects with larger quantities provides an opportunity for students to develop and deepen their understanding of unitizing, as they discover that grouping by units larger than one leads to more efficient and accurate counting.

Counting Collections

Counting Collections is a structured activity that allows children to work on all three counting principles as well as use and strengthen their developing understanding of early number concepts such as hierarchical inclusion, conservation, unitizing, and base-ten understanding, in tandem and in support of each other (Franke et al., 2018). In this activity students count different-sized collections of markers, pencils, math manipulatives, buttons, bottle caps, dice, or any countable items, to answer the question "how many are there?" As students work in pairs to count, organize, keep track of, and record their results, they develop understanding of the complex concept of number and increasingly sophisticated strategies that make sense to them.

The materials that children count can influence their success. Objects that can be lined up or easily moved will support organization and keeping track. Objects of different colors or sizes may lead students to sort and then count separate sets. The size of the set will also influence the difficulty, and can be varied to differentiate the activity. Asking questions and making observations while students are working on counting their collections can help the teacher gather formative assessment information about student thinking and understanding. Observation should include how the children are counting, how they are organizing the collection, and how they are keeping track.

 While students are counting their collections ask: How are you counting? How many do you have? How do you know? Are you sure? Is there a way to keep track of the ones you have counted? Could you count them a different way?

Counting Collections is a purposeful activity where children will use their developing knowledge of number while working on principles that have not yet been developed. Note that children do not need to learn all of these principles and concepts before engaging in this activity; rather they can be supported to learn them through the activity. It is important to allow students to develop their own system for organizing the count, so that it makes sense to them given their developing understanding, while encouraging them to develop more efficient strategies (Carpenter et al., 2017). Counting Collections is also a useful activity for children in second and third grade who are developing base-ten and multiplicative understanding.

Figure 3.15 shows a range of strategies that kindergarten children used to count and record a collection of 15 square tiles. What strategies and evidence of understanding do you see in their work?

Figure 3.15 Four children's written recording for counting a collection of 15 objects

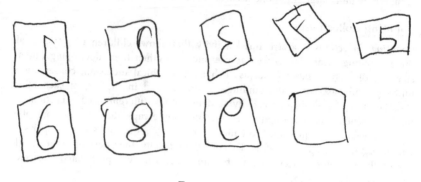

Evan

Macie

Nicole

Figure 3.15 Continued.

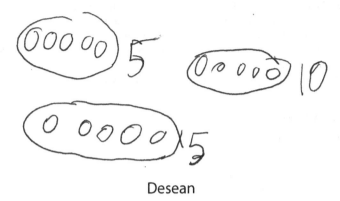

Desean

Evan shows the understanding of one-to-one correspondence, writing one numeral in each box, but is not able to represent 15. The evidence suggests that Evan may not have mastered the counting sequence past six. Macie and Nicole show the ability to represent a set of 15 objects through their drawings; Nicole shows her understanding of one-to-one correspondence while Macie shows an understanding of cardinality (using the numeral 15 to represent the quantity of the set). Desean is able to group the objects in sets of five and also count those groups by increments of five, showing an understanding of early unitizing.

Chapter Summary

- Learning to count a set of objects involves developing understanding of three principles: sequence of the number words, one-to-one correspondence, and cardinality.
- Children develop more flexible strategies for using counting to solve problems as they come to understand several interrelated and foundational concepts: hierarchical inclusion, conservation, and perceptual and conceptual subitizing.
- Once children develop an understanding of cardinality and conservation, an important milestone is achieved; they are able to develop more sophisticated and efficient problem solving strategies, such as counting on, using part–whole relationships, and counting by groups.
- Engaging young children in solving meaningful problems around determining and comparing quantities helps to build a foundation for developing a conception of number that moves beyond treating quantities as collections of ones (Nunes & Bryant, 1996).

Looking Back

1. **Counting Collections:** Counting Collections allows students to use developing understanding while working on principles not yet developed. It is important that students develop their own system of organizing when

counting so that it makes sense to them at their current level of understanding.

(a) What features of the collection of items presented below in Figure 3.16 can influence student success with counting the collection?

(b) Review the student evidence of counting in Figures 3.1 and 3.14. What strategies for organizing and keeping track might students use on the collection in Figure 3.16? Try to anticipate strategies that reflect a variety of levels of skill and understanding.

Figure 3.16 A collection of circles.

How many shapes are there?

2. **Developing Understanding of Counting Principles:** This chapter introduced the idea that in order for children to successfully count to determine the numerosity of a set they must coordinate three logical principles: knowing the counting sequence, one-to-one correspondence, and cardinality. Look again at Ethan and Hailey's work (Figure 3.5) from earlier in the chapter.

(a) What understanding of these principles is evident in the work?

(b) What understanding is still developing?

(c) Based on the evidence in Hailey's work, what might happen if she was presented with a quantity to count that is less than 15? What evidence in her work supports this thinking?

(d) What instructional next steps would you recommend for both students based on the evidence in their work?

3. **Developing More Sophisticated Strategies:** Recall that once students have developed the interrelated concepts of cardinality, hierarchical inclusion, conservation, and perceptual and conceptual subitizing they are able to understand more sophisticated strategies like counting on. Review the task and student response from Figure 3.9.

(a) What aspects of the problem and student response are related to foundational concepts: hierarchical inclusion, conservation, and cardinality?

(b) How do these interrelated ideas come together in the student's response?

(c) How does understanding foundational ideas support the use of the more sophisticated strategy of counting-on?

4. **Connecting to the OGAP Addition and Subtraction Progressions:** How do the concepts of hierarchical inclusion and conservation of number play a part in the use of strategies at the counting level of the *OGAP Addition and Subtraction Progressions*?

5. **Producing a Set:** Figure 3.17 shows four student responses to a task that asked them to build and draw a tower with 8 cubes.

Figure 3.17 Build and draw a tower with 8 cubes

Student A

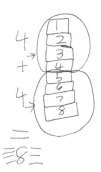

Student B

Figure 3.17 Continued.

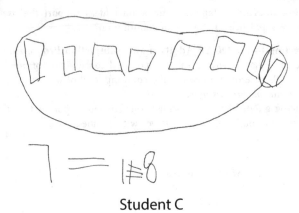

$$7 = | \not\equiv 8$$

Student C

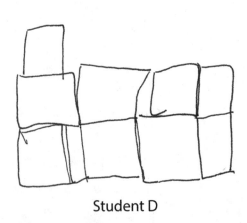

Student D

(a) What concepts are evidenced in their work?
(b) What concepts are still developing?
(c) What are appropriate next instructional steps for each student based on the evidence in their work?

Instructional Link

Use the following questions to analyze the ways your math instruction and program provide opportunities for students to develop counting and early number concepts.

1. In what ways does your math program and instruction allow students regular opportunities to count sets or collections of items?

    (a) What opportunities do students have to count a variety of sets and collections?

    (b) How do your program and instruction support students in developing strategies for organizing, keeping track, and recording the results of counts?

2. How does your math program approach foundational concepts of counting and number (hierarchical inclusion, conservation, cardinality, counting on, and perceptual and conceptual subitizing)?

    (a) Are there meaningful opportunities for determining how many and for comparing quantities?

    (b) What opportunities for subitizing are present in the program and your instruction?

4
Unitizing, Number Composition, and Base-Ten Understanding

Big Ideas

- Understanding the base-ten structure of the number system involves coordinating knowledge about the written notional system and underlying concepts of part–whole, unitizing, and multiplicative understanding.
- Children's understanding of number develops from conceptualizing quantities as collections of ones to unitizing, which allows them to make and use groups of tens to count and compare multidigit quantities.
- A further development in base-ten understanding is the ability to conceptualize and use composite units, or units made up of smaller units, without needing to see the smaller units.
- As children learn to compose and decompose numbers into place value parts, they can use that understanding to count, compare, and compute with multidigit numbers.
- Instruction that focuses only on the positional values of digits in a number and rote or procedural use of base-ten models can promote a limited understanding that focuses on the face value of the digits rather than underlying place value.

The Base-Ten Structure of the Number System

Our number system utilizes a base-ten structure that allows for a great degree of efficiency in both notation and computation. For example, to write the number 3768 we use just 4 digits. In contrast, to write the same number the Romans had to use a string of 12 letters, using combinations of units of 1000 (M), 500 (D), 100 (C), 50 (L), 10 (X), 5 (V), and 1 (I) to compose the number additively as MMMDCCLXVIII. The efficiency of the base-ten number system is based on the use of 10 digits (0 to 9), and four properties (Ross, 1989):

1) The *positional* property: The position of a digit determines its value in units (e.g., the digit 2 can represent 2 ones, 2 tens, 2 hundreds, 2 tenths, etc., depending on its position in the numeral).

2) The *base-ten* property: The value of every position is ten times the value of the position to its right and 1/10 the value of the position to its left (e.g., 100 is 10 times 10, 1 is 1/10 of 10).

3) The *multiplicative* property: Multiplying any digit in the numeral by the place value of the position determines the actual value of each digit (e.g., in 532, 5 × 100 is the value of the 5, 3 × 10 is the value of the 3, and 2 × 1 is the value of the 2).

4) The *additive* property: The value of the entire numeral is equal to the sum of the values of each of the digits (e.g., 532 = 500 + 30 + 2).

The combination of these four properties is what makes our number system so compact and useful for arithmetic. Most algorithms for computation have their foundations in the base-ten property or the idea that ten units can be regrouped into a unit of the next size. This makes it possible to carry out a computation like 5994 + 3289 as a series of single-digit additions in each place value. In this way our number system helps to off-load some of the cognitive effort of computation onto the written notation (Nunes & Bryant, 1996). Imagine trying to add 5,994 + 3289 in Roman numerals: MDCMXCIV + MMMCCLXXXIX.

Understanding that 5994 is close to 6000 makes it into an even easier problem to solve. Taking 6 from 3289 and adding it to 5994 turns the problem into 6000 + 3283, or 9283.

5994 + 6 = 6000
3289 − 6 = 3283

This kind of fluency and flexibility with computation relies on an understanding of the structure of the number system, number relationships, and properties of operations (Russell, 2000). Understanding the structure of the base-ten number system can also lead to a deeper understanding of magnitude and estimation, as well as lay the foundation for algebra (Howe & Epp, 2008).

Chapter 6 Developing Whole Number Addition and Chapter 7 Developing Whole Number Subtraction for more on computational fluency and algorithms.

This chapter focuses on how students develop an understanding of the base-ten structure of our number system, a concept commonly referred to in mathematics classrooms as place value. This understanding is much more complex and multifaceted than it may first seem. As Ross (1989) states, it is an extension of *part–whole understanding*, discussed in Chapter 3, which allows students to decompose and compose numbers by place value parts, connecting the written numeral to the quantity it represents:

A student who understands place value knows not only that the numeral 52 can be used to represent "how many" for a collection of 52 objects but also that the digit on the right represents two of them, the digit on the left represents fifty of them (five sets of ten), and that 52 is the sum of the quantities represented by the individual digits.

(p. 47)

How does this sophisticated part-whole understanding develop? At the heart of understanding the base-ten number system is the ability to *unitize* and use groups of ten and then later hundreds, thousands, and eventually all the place value units, including decimals.

 Chapter 3 The Development of Counting and Early Number Concepts for more on part-whole understanding and unitizing.

Early Unitizing with Models

The first significant development on the path to base-ten understanding is evident when children can understand that a group of ten ones can also be conceptualized as one unit of ten. This understanding allows children to count, compare, and operate by units of ten in addition to units of one. Children can often first unitize by smaller numbers like 2 and 5 but unitizing by 10 is a critical foundation for understanding place value and therefore an important focus of early math instruction.

The examples of student thinking shown in Figure 4.1 illustrate a range of student strategies for determining the total shown on two sets of hands. In this example, the objects to be counted (fingers) are shown as single units but are also grouped into fives and tens. Whether or not students use those groups to determine the total provides insight into their developing understanding.

Figure 4.1 Mason, Evelyn, and Arya's strategies for counting quantities grouped by five and ten

Two children made a number together by holding up fingers. What number did they make together?

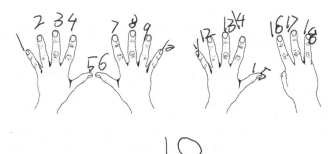

Mason's strategy

Figure 4.1 Continued.

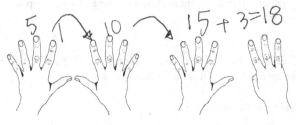

Evelyn's strategy

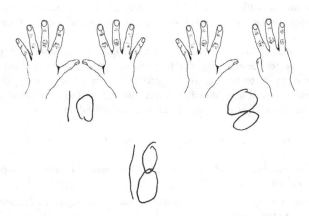

Arya's strategy

Mason determines the total by counting each finger one by one. Evelyn's work shows the ability to unitize by skip counting each full hand by 5's and then adding on 3 from the last hand. Arya recognizes that 10 fingers or two hands together make one 10 and that the second set of hands shows 8. Arya's strategy for determining the total shows the connection to the place value of 18 as 10 and 8. The evidence in Evelyn and Arya's work shows the ability to unitize when the ones are visible and are examples of *Early Transitional Strategies* on the *OGAP Base Ten Number Progression*.

Recognizing that multidigit numbers are made up of tens and ones is made more difficult by the spoken form of numbers, particularly those between 10 and 20, or the teen numbers. There is not a direct connection between the word eighteen and "ten and eight" and this connection is even more obscure for the numbers 11, 12, 13, and 15. After 20, the place value connection to spoken words is somewhat more straightforward in that the number of tens is said first and then the number of ones; however the decade names do not translate directly into the number of tens: it is not obvious that twenty means "two tens." Children need frequent and diverse opportunities to interact with multi-digit quantities and both visual and concrete models in order to make the connection to both the written notation and spoken form. Models such as hands, ten-frames, and structured bead strings that show both single units and groups of five or ten are important for supporting the development of this concept.

 Chapter 5 Visual Models to Support Additive Reasoning for more on visual models.

In the example shown in Figure 4.2, Claire fills in ten-frames and keeps track of the running total of full ten-frames to compose 38 with 4 ten-frames.

Figure 4.2 Claire's strategy for making the number 38 in ten frames

If you make 38 with tens frames, what is the least number of tens frames you can use.

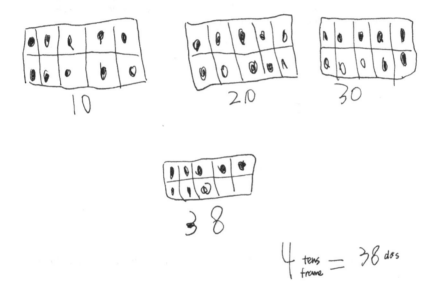

An important follow-up question for her teacher would be to find out if she understands that 20 is the total number of dots in two full ten-frames rather than being a label for the second ten-frame. This is, in many ways, parallel to children's understanding of cardinality when counting by ones, or understanding that the last counting word signifies the whole quantity.

 When students count a collection by tens, ask them what the 20 (or 30, 40, 50, etc.) represents. If they point to the 2nd (or 3rd, 4th, 5th, etc.) group of ten rather than the number of groups of ten, this may be an indication that they have a rote understanding of counting by tens.

Developing Abstract Composite Units

A second significant milestone occurs when children can unitize in tens without need-ing to see or draw the ones and can mentally anticipate and use the result of that unit-izing. This development does not happen all at once, but rather proceeds in phases and depends upon the context or situation and the size of the numbers.

The task in Figure 4.3 involves figuring out how many groups of ten are in 56. Alyssa begins to unitize by tens without needing a model, drawing on a known fact that 10 + 10 = 20, but then begins to draw out each group and count by ones. She also loses sight of the groups she has made and answers the question with the total rather than the number of groups. Her work is a good example of some of the chal-lenges children may have in transitioning from units of one to units of ten.

Figure 4.3 Alyssa adds by tens but then counts by ones to determine the number of tens in a two-digit quantity

> Shonda bought packs of glow bracelets to give out to all the kids in her grade. Each pack has 10 glow bracelets in it. There are 56 kids in Shonda's grade. How many packs of bracelets did she need to buy?

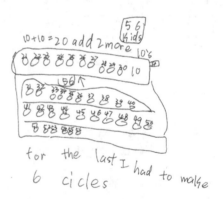

In Figure 4.4, Chantelle solves the same problem by showing the ability to make groups of ten and also count by tens.

Figure 4.4 Chantelle makes groups of ten and counts by tens.

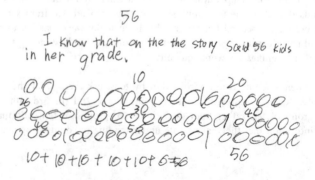

Chantelle's work shows evidence of understanding ten as a countable unit, but she has to keep counting out and drawing a new group of ten every time she adds another group of ten. Her work does not show the ability to conserve or anticipate the result of making a group of ten. In many ways this is similar to children who don't have *conservation* yet and have to recount quantities from one.

Go To Chapter 3 The Development of Counting and Early Number for more on conservation.

In contrast, Brock's work in Figure 4.5 shows that he can work abstractly with units of ten, counting them up until he reaches 56. However, like Alyssa in Figure 4.3, Brock gets confused when it comes time to answer the question, incorrectly answering that Shonda will need 50 packs rather than 5.

Figure 4.5 Brock counts by units of ten without a model

Shonda bought packs of glow bracelets to give out to all the kids in her grade. Each pack has 10 glow bracelets in it. There are 56 kids in Shonda's grade. How many packs of bracelets did she need to buy?

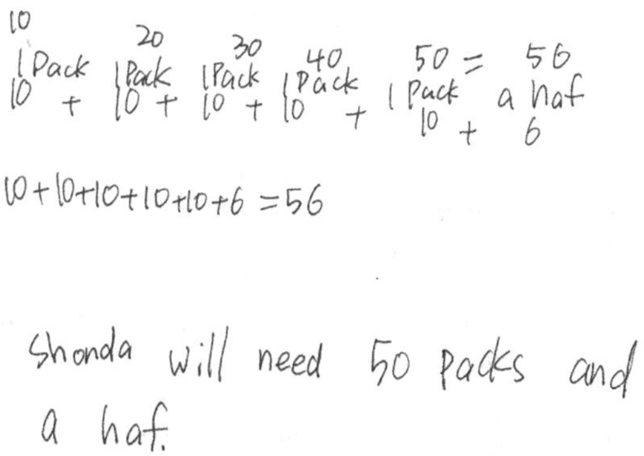

$$10$$

$$\underset{10}{1\ Pack} \underset{+}{\overset{20}{1\ Pack}} \underset{10}{\overset{30}{1\ Pack}} \underset{+}{\overset{40}{1\ Pack}} \underset{10}{\overset{50}{1\ Pack}} \overset{56}{=} \underset{a\ haf}{6}$$

$$10 + 10 + 10 + 10 + 10 + 6 = 56$$

Shonda will need 50 packs and a haf.

Despite his error, Brock's strategy shows significant progress towards fluency with base ten in that he does not need to actually model or group the bracelets into packs of ten as Alyssa (Figure 4.3) or Chantelle (Figure 4.4) did, but rather can mentally anticipate the result of grouping them by repeatedly adding and counting by tens. He does not yet show the understanding that 56 is composed of 50 and 6, or 5 tens and 6 ones, but he is able to use tens to increment up to 56. For him, ten is a *countable unit* (Cobb & Wheatley, 1988). His strategy falls in the *Transitional* level of the *OGAP Base Ten Number Progression* because he is unitizing by composite units without needing to see the ones.

It is important to note that it is easier for children to count tens and then ones than it is for them to start with a number of ones and count on that number by tens (e.g., 3, 13, 23, 33, 43 …). Counting off the decade involves simultaneously keeping track of the tens and ones. In Figure 4.6, both students show an understanding of relative magnitude in identifying the first number counted that is greater than 75. However, while Lizzy shows the ability to count fluently by tens starting at 43, Enakshi does not.

Figure 4.6 Lizzy and Enakshi's responses to a task involving counting by tens off the decade

Count by tens starting at 43. What is the first number you say that is larger than 75?

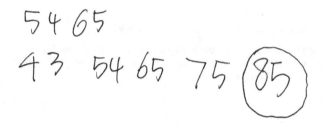

Lizzy's solution

$$54 \; 65$$

$$43 \quad 54 \quad 65 \quad 75 \quad \boxed{85}$$

Enakshi's solution

Enakshi's solution shows that she understands there is a pattern of the tens digit increasing when counting by tens, but she also initially increases the ones digit. Once she gets to 65, she is able to count correctly by tens, perhaps because counting by tens off a multiple of 5 is more familiar. Children need opportunities to count by tens, both on and off the decade, using visual models, as this will become an important strategy for both addition and subtraction.

 Chapters 6 Developing Whole Number Addition and Chapter 7 Developing Whole Number Subtraction for the importance of counting by tens in addition and subtraction.

Understanding Numbers as Composed of Place Value Parts

The ability to understand and work with composite units sets the foundation for a third milestone: understanding how numbers are composed of units of different sizes. A multi-digit number like 48 can be composed in many ways: 45 + 3, 20 + 28, etc., but the decomposition into 40 + 8 uses units of tens and ones and is directly linked to the written numeral. This composition by place value parts can be understood additively, as levels of units, or multiplicatively, as shown for the two- and three-digit numbers in Table 4.1.

In the student solutions in Figure 4.7, Zack shows an understanding of 56 as the sum of 50 and 6 and 60 as composed of 6 tens, but then does not express his answer in units of ten, resulting in an incorrect answer (60 packs). Cena's solution illustrates that she has a mental model of two-digit numbers that incorporates two levels of units—when the ones are not visible—and that she can keep track of and

Table 4.1 Different ways of understanding the composition of multi-digit numbers

Number Composition	48	632
Additive	40 + 8	600 + 30 + 2
Levels of units	4 tens and 8 ones	6 hundreds, 3 tens, 2 ones
Multiplicative	$(4 \times 10) + (8 \times 1)$	$(6 \times 100) + (3 \times 10) + (2 \times 1)$

distinguish these units. (The 6 is 6 packs or 6 tens.) Cena's depiction of the number line shows the ability to think about multiples of ten as units of ten, allowing her to correctly answer the question as 6 packs.

Figure 4.7 Zack and Cena's solutions show evidence of working with different levels of units

Shonda bought packs of glow bracelets to give out to all the kids in her grade. Each pack has 10 glow bracelets in it. There are 56 kids in Shonda's grade. How many packs of bracelets did she need to buy?

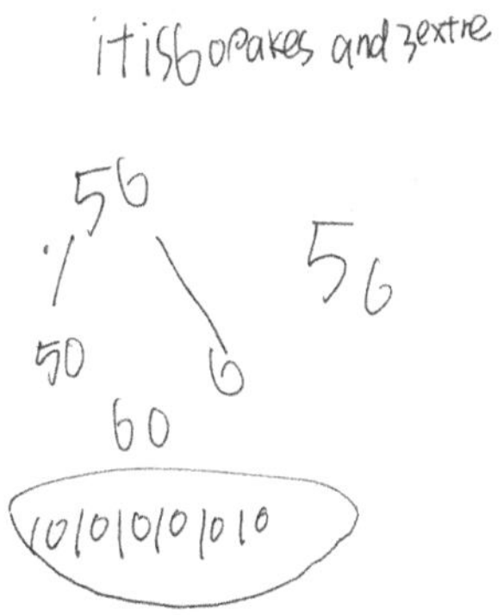

Zack's solution

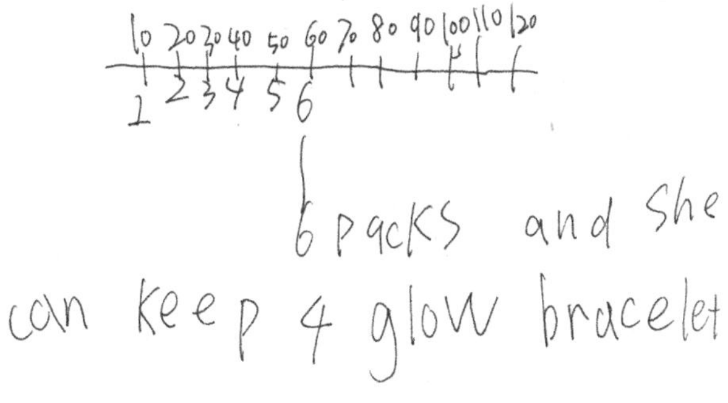

Cena's solution

Ultimately, it is important for students to develop both of these understandings—composing and decomposing a number by place value parts and conceptualizing a number in terms of two or more levels of units.

With three-digit numbers, number composition extends to an additional level of unit: one hundred can be represented as one group of 100, 10 groups of 10, or 100 individual units. Understanding that 10 tens is equal to 100 leads to the understanding that 30 tens is 3 hundreds or 300. In the task shown in Figures 4.8 and 4.9, students are asked to write the number that is composed of 20 tens and 9 ones. Their responses to this question provide insight into their understanding of number composition with three-digit quantities.

Figure 4.8 Jaden's and Leon's responses show evidence of counting by units of ten

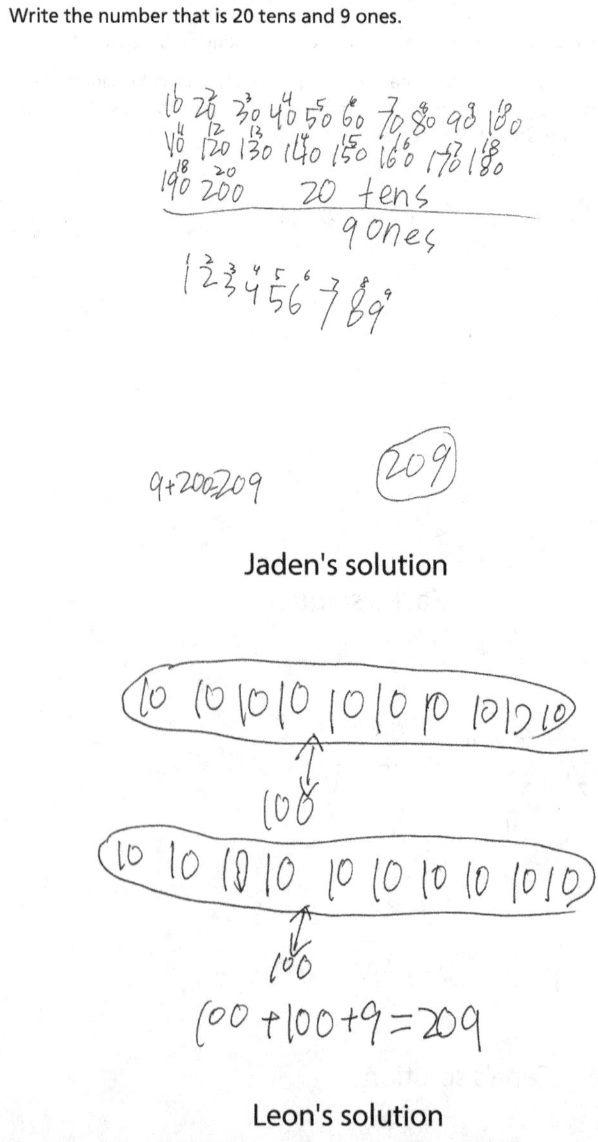

Write the number that is 20 tens and 9 ones.

Jaden's solution

Leon's solution

In Figure 4.8, Jaden skip counts by tens, keeping track of the number of tens to get to 20 tens or 200 and then counting out the ones to determine the total is 209. Leon writes out 20 tens, but arranges them into groups of ten, showing the understanding that one hundred, or 100, is a unit of made up of 10 tens. Both strategies are considered *Transitional* on the *OGAP Base Ten Number Progression*.

Natalia's solution, shown in Figure 4.9, shows evidence that she recognizes that 200 is composed of 20 tens without needing to show or count the individual tens. Her strategy reflects understanding at the *Base-Ten* level of the progression, *Number Composition by Place Value Parts* (hundreds, tens, and ones).

Figure 4.9 Natalia's solution shows evidence of number composition by units of ten

Write the number that is 20 tens and 9 ones.

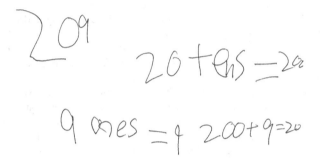

This progression from unitizing to composite units to number composition is also reflected in the CCSSM for Number and Operations in Base Ten in grades K–3.

CCSSM on the Development of Base-Ten Understanding

The CCSSM lays out a progression in grades K–5 of expectations around developing understanding of base ten and using this understanding to compose and decompose numbers.

- In kindergarten, children develop the understanding that teen numbers are composed of "ten ones and some further ones," or unitizing when the ones are visible.
- In first grade, this extends to understanding that "10 can be thought of as a bundle of ten ones," understanding teen numbers as composed of a ten and a number of ones, and understanding multiples of ten as units of tens (e.g., 30 is 3 tens). This transition from understanding ten as a group of ten ones in kindergarten to unitizing groups of ten in first grade is subtle but significant.
- Second grade builds up to unitizing by composite units to understand and apply the concept of 100 as "a bundle of ten tens" and understand multiples of 100 as units of hundreds (one hundred, two hundreds, etc.).

- In second grade students are also expected to use this understanding of composite units to represent, compose, and decompose three-digit numbers into place value parts (as sets of hundreds, tens, and ones).
- In third grade, where the focus shifts to multiplication, students should use place value understanding to see patterns when multiplying by multiples of ten.
- In fourth grade, students should be able to use a multiplicative understanding of place value, understanding that each place value is ten times the place value to its right and using expanded form to represent multi-digit numbers, e.g., 426 is (4 × 100) + (2 × 10) + (6 × 1).
- If fifth grade, this understanding should extend to decimal numbers.

Face Value without Base-Ten Understanding

Several research studies have shown that children may develop an understanding of the positional value of digits in a number without the underlying understanding of unitizing by composite units (Kamii, 1982; Ross, 1989; Sinclair, Mello, & Siegrist, 1988, as cited in Nunes & Bryant, 1996). Student work samples collected through the *OGAP* project also reveal this misconception. The task shown in Figure 4.10 was designed to elicit evidence of base-ten understanding by asking how many tens are in a three-digit number. In Figure 4.10, Mateo and Yasmeen answer the problem correctly by using their developing base-ten understanding. Mateo counts up by units of ten to prove that there are 18 tens—a *Transitional* strategy—while Yasmeen uses a strategy that shows understanding of *Number Composition by Place Value Parts* (100 is made up of 10 tens and 80 is made up of 8 tens).

Figure 4.10 Mateo and Yasmeen's solutions show evidence of developing base-ten understanding

Tanya said there are 18 tens in the number 182. Is she right or wrong?

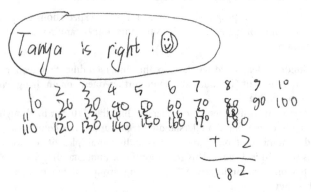

Mateo's solution

Figure 4.10 Continued.

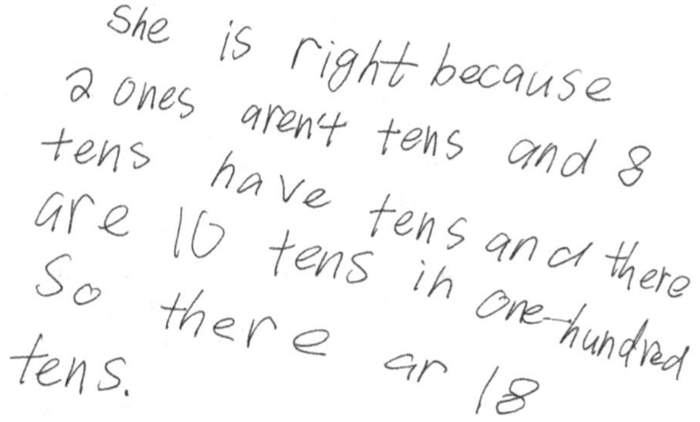

She is right because
2 ones aren't tens and 8
tens have tens
are 10 tens and there
So tens in one-hundred
tens. there ar 18

Yasmeen's solution

Michaela's incorrect solution in Figure 4.11 shows that she can represent the number of hundreds, tens, and ones in the number 182 using the corresponding base-ten blocks. However, although she knows what the hundred block looks like, the evidence does not show an understanding that it is made up of 10 tens blocks. Michaela directly models each place in the number with the correct size unit but does not show understanding of the ten-to-one relationship between these units. She knows there are 8 tens in 182, but the place value chart she uses does not help her think about the hundred as being composed of 10 tens. Michaela's representations of these blocks reflect the relative size of the units but not the ten-to-one relationship.

Figure 4.11 Michaela directly models the value of each digit in a three-digit number

Tanya said there are 18 tens in the number 182. Is she right or wrong?

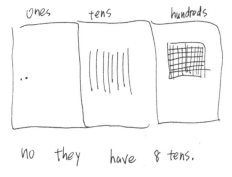

no they have 8 tens.

Some researchers argue that this incomplete understanding may be the direct result of math instruction that focuses almost exclusively on the positional property of the base-ten number system (Cobb & Wheatley, 1988; Kamii & Dominick, 1998). Questions typically seen in curriculum materials to assess students' understanding of place value such as, "How many tens are in 73?" or "Which digit is in the tens place in 576?" focus students on the *face value* of the digits (7) rather than the actual value (70). In such cases, students may think of 7 tens as a unit label in the same way they think about 7 balloons or 7 cents, but not necessarily as composite unit equivalent to 7 groups of 10 ones. The fact that students can answer these types of problems successfully may lead teachers to conclude that they understand place value. Even when textbooks include visuals of sticks bundled in tens or base-ten blocks, students may treat these collections as if they were just different-sized units rather than being composed of units (i.e., the ten block is composed of ten ones) (Cobb & Wheatley, 1988). These misconceptions come to light when asked a question like, "how many tens are there in 327?" Students who answer 2 may be focusing on face value over place value.

Even more troubling is the fact that many students can progress through elementary school mathematics without ever developing this base-ten understanding. Ross (1989) found that the majority of students in grades 2 through 5 that she interviewed were unsuccessful in solving a task that involved identifying the number of objects that were represented by the tens and ones digits in a collection of 25 objects. Less than half of the students could successfully identify that the 2 in 25 represented 20 objects. However, when she presented the same task with 52 represented in base-ten blocks (5 ten blocks and 2 ones blocks) some of those same students were successful because they could rely on the "face value" of the digits to answer correctly. In essence, to them the 5 meant 5 of something (tens blocks) and the 2 meant two of something else (ones blocks), but they did not see that the 5 meant 5 groups of ten ones. More concerning, she found that even many fourth and fifth grade students still had a face-value understanding of the digits in multi-digit numbers.

 To learn more about a student's understanding of place value, ask them to count out a number of objects greater than 10 and write the number. Point to the digit in the ones place and ask them to show you with the objects what that digit means. Then point to the digit in the tens place and ask the same question. See Figure 4.12.

Figure 4.12 Eliciting student understanding of place value

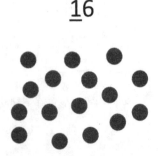

See Looking Back Question 2 (Figure 4.18) for examples of students' written responses to this question.

Figure 4.13 shows another example of a student who is still developing understanding of composite units. Jared answers the problem about packs of 10 in 56 by drawing out 5 tens and 6 ones and counting by tens, but his answer to the problem is 11 packs. Although he counts each pack by ten, the evidence does not suggest an understanding that the units of ten are different from the units of one and counts them all up as if they were all the same size units. He also refers to the leftovers as packs. While there is some developing understanding to build on here, Jared does not show an understanding of ten as a composite unit made up of ten ones in order to successfully solve this problem.

Figure 4.13 Jared models 56 correctly in tens and ones but disregards the difference in units

Shonda bought packs of glow bracelets to give out to all the kids in her grade. Each pack has 10 glow bracelets in it. There are 56 kids in Shonda's grade. How many packs of bracelets did she need to buy?

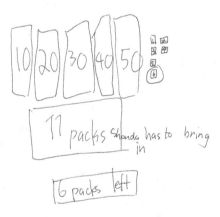

A focus on the face value of multi-digit numbers without underlying base-ten understanding can also lead to students making conceptual errors in computation such as the example in Figure 4.14.

Figure 4.14 A face-value approach to multi-digit addition leads to an unreasonable answer

The red ribbon is 79 inches long. The red ribbon is 83 inches shorter than the yellow ribbon. How long is the yellow ribbon?

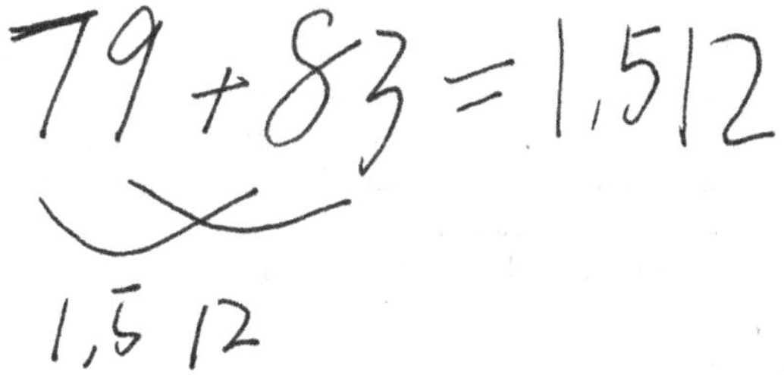

$$79 + 83 = 1,512$$

1,5 12

To add 79 and 83, this student added the digits in the tens place (7 + 8 = 15) and the digits in the ones place (9 + 3 = 12) and put those digits together to get an answer of 1512. The student added the ones and tens separately but added the face values rather than the actual values of the digits. The fact that the student did not recognize that an answer in the thousands was unreasonable is further evidence that there is a need to develop base-ten understanding. As discussed in Chapters 6 and 7, and in the CCSSM, strategies for addition and subtraction, as well as multiplication and division, should be based on an understanding of place value and properties of operations.

Fortunately, these misconceptions are not inevitable. Building strong understanding of base-ten number in the elementary grades provides the foundation for the development of efficient and flexible computational strategies, understanding of relative magnitude, and understanding of decimal numbers. As discussed in Chapter 5, visual models that show the ten-to-one relationship can be an important tool for developing this understanding. Another important way to develop this foundation is by giving students multiple and varied opportunities to use counting to solve problems with multi-digit numbers, share their solutions, and develop more efficient strategies.

 Go To Chapter 5 The Role of Visual Models in Additive Reasoning for more on using visual models to support the ten-to-one relationship.

Instructional Strategies to Build Base-Ten Understanding

In Chapter 3, we discussed an activity called Counting Collections, where children count collections of objects to answer the question "how many are there?" (Franke, Kazemi, & Turrou, 2018; Schwerdtfeger & Chan, 2007). By giving students the opportunity to count larger collections, they will naturally see the need to organize the objects in groups in order to count efficiently and keep track. Putting 10 objects in a pile, a cup, or in a ten-frame can be a helpful organizational tool for large quantities and help students see the connection between quantity, the written numeral, and the composition of place value parts. The following vignette illustrates how the Counting Collections activity can be facilitated to focus on developing base-ten understanding.

Counting Collections in Second Grade

Ms. Milewski, a second grade teacher, had her students count collections at various points throughout the year, beginning with quantities less than 100. In February, she created bags of objects like bottle caps, square tiles, and buttons with between 100 and 300 objects. Before sending students off to count, she led a whole group discussion about strategies that would help them count efficiently.

Ms. Milewski:	When we did Counting Collections before, we talked about strategies that are efficient. This time your collections are bigger, so now you really need to think about strategies you are using when counting. Who can explain what efficient strategies are?
Jesse:	Counting by fives or tens.
Ms. Milewski:	Ok, can you explain that a little more?

Jesse:	You take ten and then you keep making ten and count out those.
Ms. Milewski:	Who can tell me why that would be an efficient strategy?
Shanice:	Because if you are doing something else like color, then you will have a lot of different numbers to add, and if you are doing by fives or tens you can just count by tens.
Ms. Milewski:	Does anyone want to add on?
Ebony:	Because counting by ones you have to start over, and you might miss some and get a different answer.
Ms. Milewski:	We found that out with our first Counting Collections, didn't we? Some people went back to check and see if their answer was right, and it usually wasn't the same answer when they were counting by ones.
Akeno:	And it's going to take longer.

After this discussion and a review of the norms and expectations for the activity, Ms. Milewski paired students strategically and sent them off to count collections and record their counts on a recording sheet. Figure 4.15 shows two students counting a collection of bottle caps by making piles of 10 and their written recording. The evidence in their work shows that they determined the total by counting by tens and then ones to get a total of 268.

Figure 4.15 Two students count and record their collection by making piles of 10 and counting by 10

Figure 4.15 Continued.

Their work shows that they were able to treat each pile of 10 as a countable unit and could count up by tens to 260 and then by ones to get a total of 268. Ms. Milewski noticed that there were several pairs of students who could make and use groups of 10 or 20 to find the total, but also that no one was making those groups into larger groups of 100.

After the students in the classroom had engaged in counting collections of different sizes, Ms. Milewski pulled the class together for a discussion. Since many students had counted by tens, she asked several students to report out on how many piles of 10 they had made, how many were left over, and the total amount. Together the class made the chart shown in Figure 4.16.

Figure 4.16 Chart of class discussion of tens and ones in their students' collections

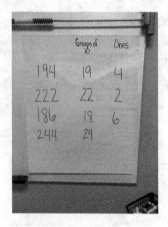

After collecting three different examples, students began to see a pattern between the digits in the numeral and the groups of tens and ones. When one group shared that they had counted 244 and 24 groups of ten, Ms. Milewski asked the class how many they thought were left over. When students were able to correctly "guess" 4 she asked them what patterns they were noticing and emphasized the place value of the numbers.

Ms. Milewski:	Can anybody tell me what you noticed? About the groups of ten that you made and the number? Turn and talk about that with your neighbor. What pattern do you notice?
Jesse:	I noticed that all the groups of ten are the numbers at the beginning of the number. So, like 19 and 4 makes 194
Ms. Milewski:	So, the 19 is 19 groups of ten and it's also the first two digits of the total. Anybody else?
Gabriel:	The first two numbers are going to be how many tens and the last number is going to tell you how many left over.
Ms. Milewski:	So, if we made the number 568, how many groups of tens do you think we would make? How do you know? Turn and talk.

Although many students had made groups of 10 to count their collections, they hadn't seen this connection between the groups of ten and the written numeral until the whole group sharing and discussion. Ms. Milewski used the activity of Counting Collections to support the movement of strategies on the *OGAP Base Ten Number Progression*. She also learned from this activity that an important goal for instruction was to help students understand that a group of 10 tens could be thought of as one group of 100.

The Application of Base-Ten Understanding

Once students have developed strong and flexible understanding of base ten and number composition, they can apply that understanding to a wide range of tasks and content areas. At the top of the *OGAP Base Ten Number Progression* is a level called *Application of Base Ten*. This level involves being able to flexibly use base-ten understanding to solve problems, develop efficient computation strategies, and also extend this understanding to decimal place values and negative numbers.

CCSSM on the Application of Base-Ten Understanding

The CCSSM expectations for grades 1–5 reflect the application of developing base-ten understanding to compare and perform operations with multi-digit numbers, as shown in Table 4.2.

- In grades 1–3, students are expected to apply their understanding of unitizing, composite units, and place value to develop strategies for adding, subtracting, rounding, and comparing single and multi-digit numbers.
- In grade 3 through 5 they are expected to apply base-ten understanding to develop and refine strategies for multiplication and division.

Table 4.2 Common Core expectations for application of base-ten understanding

Grade	Understanding	Application
1	Unitizing by tens and composing/decomposing teen numbers by tens and ones	• Add and subtract single-digit numbers or multiples of 10 to 2-digit numbers • Compare 2-digit numbers
2	Unitizing by hundreds and composing/decomposing 2-and 3-digit numbers into place value parts	• Add and subtract 2-digit numbers • Compare 3-digit numbers
3	Base-ten understanding within 1000	• Add, subtract, and round 3-digit numbers • Multiply by multiples of 10
4	Generalize base-ten understanding to 1,000,000 and to decimals in the tenths and hundredths	• Multiply, divide, compare and round multi-digit numbers
5	Extend understanding of the base-ten system to decimals in the thousandths	• Add, subtract, multiply, divide, and compare decimals • Use exponents to denote powers of 10 • Explain patterns when a decimal number is multiplied of divided by a power of 10
6–8	Understand the base-ten system in terms of powers of 10 and extend the base-ten system to negative numbers	• Add, subtract, multiply, divide, and compare negative numbers • Express very large or small numbers in scientific notation and express how many times larger one is than the other

- In grade 4, students generalize their place value understanding to recognize and use the multiplicative relationships in the base-ten number system by understanding that each place value increases in value by a factor of 10. For example, 40×10 is $(4 \times 10) \times 10$ or $4 \times (10 \times 10)$. Since 10×10 is 100, this means it will be equivalent to 4×100 or 400.
- In grade 5 they extend this understanding to computation with decimal numbers and understanding why multiplying and dividing by powers of ten shifts the digits a corresponding number of digits to the left or right. For more on these concepts, see *A Focus on Multiplication and Division* (Hulbert, Petit, Ebby, Cunningham, & Laird, 2017).
- In grades 6–8, students are expected to apply their base-ten understanding to work flexibly with all the numbers that can be represented on a number line, including negative numbers and irrational numbers. They are also expected to understand very large and very small numbers in scientific notation, as the product of a whole number and a power of ten (e.g., 3 billion or 3,000,000,000 is 3×10^9).

Summary

Children's understanding of number begins with counting and culminates with a deep and flexible understanding of the base-ten number system which allows them to understand relative magnitude, develop computational fluency, and extend their understanding to all rational and irrational numbers. The first major development occurs when children can unitize or simultaneously treat a group of 10 ones as one unit of 10. A further development occurs when they can conceptualize 10 as a unit that can be counted without needing to see the ones. At the next level, they can compose and decompose numbers by tens and ones in multiple ways. They will progress through these same conceptual milestones with larger and smaller units, (100s, 1000s, tenths etc.) until they will learn to generalize and extend the ten-to-one relationships to all numbers that can be represented on a number line (the set of real numbers).

Looking Back

1. **Counting by Ten Off the Decade:** Examine Elise's solution to an addition problem in Figure 4.17.

Figure 4.17 Elise's response to the baseball card problem

Hakim had a collection of 92 baseball cards. He got 53 new baseball cards for his birthday. How many baseball cards does Hakim now have?

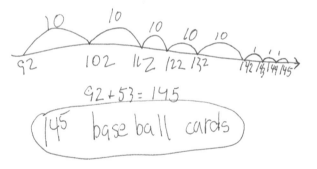

(a) What does Elise understand about number and base ten? Where does she make an error?

(b) Why is counting off the decade difficult for some students?

2. **Base-Ten Understanding and Student Solutions:** During the first week of school, Ms. Smith wanted to learn about the base-ten understanding of her new group of students. Figure 4.18 shows one of the tasks Ms. Smith used to gather evidence along with three student responses that typify the range of understanding in her class.

Figure 4.18 Three student responses showing developing base-ten understanding

Below there are 16 dots. Circle the number of dots that show what the 1 means in the number 16.

16

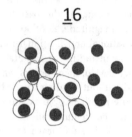

Explain your answer.

Ten Ones

Student A

16

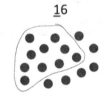

Explain your answer.

I circled 1 group of ten.

Student B

16

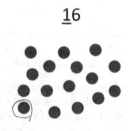

Explain your answer.

I new theat

Student C

(a) Where would each piece of student work be found on the *OGAP Base Ten Number Progression*?

(b) How is the understanding seen in student A similar to student B? How is their understanding different?

(c) What instructional next steps might be taken with a small group of students whose understanding is similar to that seen in student C?

3. **Unitizing by Composite Units.** A significant milestone occurs when children can unitize in tens without needing to see or draw in the ones and can mentally anticipate and use the result of that unitizing. Consider the student work from the Counting Collections activity in Figure 4.19.

Figure 4.19 Student recordings of Counting Collections

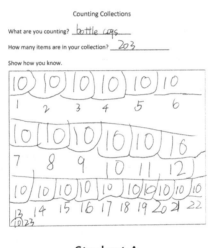

Student A

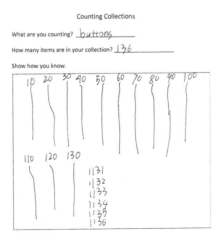

Student B

(a) What evidence indicates the students are able to treat each pile of ten as a countable unit (or composite unit)?

(b) What issues or errors are evident in the student work that need to be addressed?

4. **Number Composition by Place Value Parts:** It is important for students to develop both an ability to compose and decompose by place value parts as well as conceptualize a number as levels of units with a ten-to-one relationship (see Table 4.1.) The following questions and examples of student work all relate to these two important ideas.

The apple problem in Figure 4.20 allows students an opportunity to consider place value parts and the levels of units with ten-to-one relationships in a three-digit number. Use the problem and student solutions to consider the following questions:

(a) What ability to compose and decompose by place value parts is evidenced in the work?

(b) Beyond place value parts, what evidence is there that the students understand the composition of a number as levels of units with a ten-to-one relationship?

(c) Where would each piece of student work be found on the *OGAP Base Ten Number Progression*?

Figure 4.20 Two student responses to the apple problem

Ms. Luo's class went apple picking. The class picked 327 apples and put the apples in bags to take them home. If each bag can hold 10 apples, what is the fewest number of bags they need to take them home?

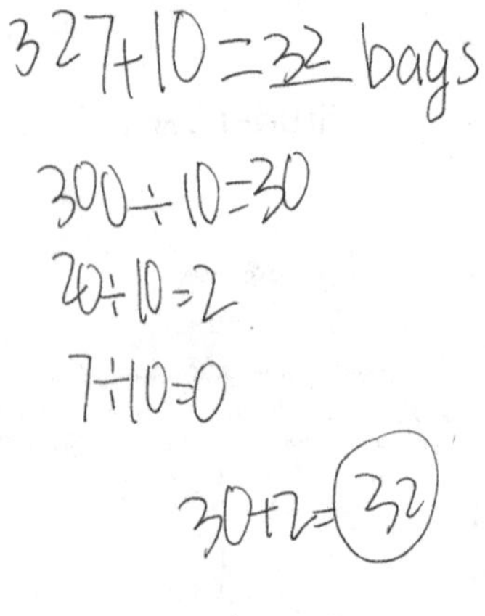

$$327 \div 10 = 32 \text{ bags}$$
$$300 \div 10 = 30$$
$$20 \div 10 = 2$$
$$7 \div 10 = 0$$
$$30 + 2 = \boxed{32}$$

Student A

Figure 4.20 Continued.

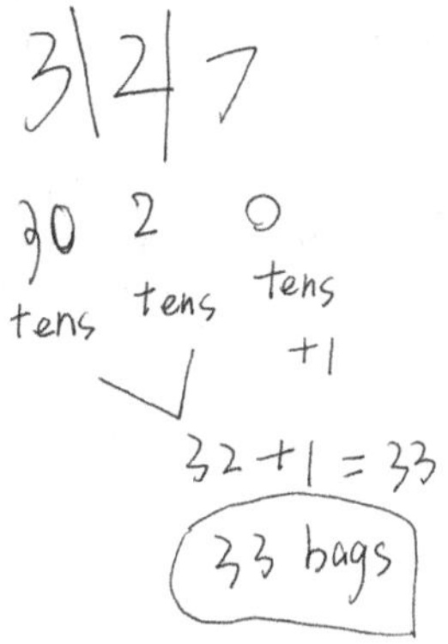

$$3\backslash2\vert7$$

$$\cancel{3}0 \quad 2 \quad O$$

tens $\quad$ tens $\quad$ tens

$+1$

$$32+1=33$$

$$\boxed{33 \text{ bags}}$$

Student B

(d) With three-digit numbers, unitizing extends to an additional level of unit: one hundred can be represented as one group of 100, 10 groups of 10, or 100 individual units. In Figure 4.21, what evidence can be seen of this understanding? What can be said about this student's ability to compose and decompose by place value parts?

Figure 4.21 A student response to a collection of pennies problem

> Pablo and Imani counted collections of pennies. Pablo has 2 piles of 100, 6 piles of 10, and 3 extra pennies. Imani has 32 piles of 10 and 5 extra. Who has more pennies?

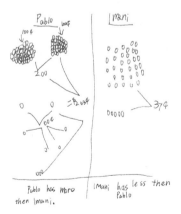

5. **Comparing and Base-Ten Understanding:** Visual models that show the ten-to-one relationship can be an important tool for developing base-ten understanding. Another effective way to develop this foundation is by giving students multiple and varied opportunities to use counting to solve problems with multi-digit numbers. Both of these elements are reflected in the rock collection problem in Figure 4.22. Use the *OGAP Base Ten Number Progression* and ideas related to relative magnitude as discussed to analyze the student work shown in Figure 4.22.

(a) How does the visual model given in the task support students in unitizing?

(b) What strategies do students use to compare the quantities?

(c) Where are the strategies the students use represented on the progression?

Figure 4.22 Four student responses to the rock collection problem

Jada and Nathan both have rock collections. Who has the bigger collection?

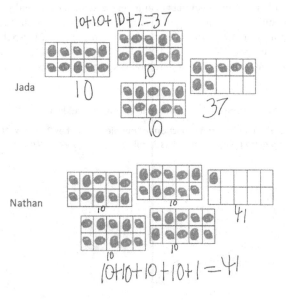

Bennett's response

Figure 4.22 Continued.

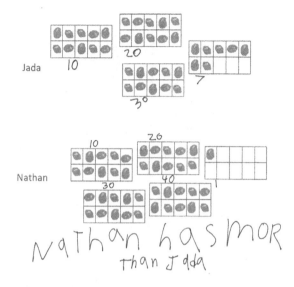

Issa's response

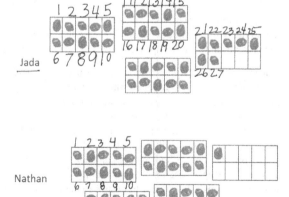

Logan's response

Figure 4.22 Continued.

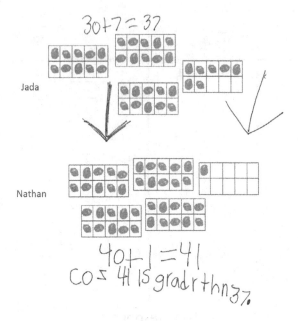

Jada

Nathan

Millie's response

Instructional Link

Use the following questions to help you consider ways your instruction and math program provide students opportunities to develop base-ten understanding through instructional emphasis on unitizing, composite units, and number composition.

1. To what degree does your instruction and math program focus on unitizing, number composition, and base-ten understanding?
2. Is systematic and intentional instructional attention given to help students develop base-ten understanding and number relationships?
3. Does the program include questions designed to build more than just face value understanding of multi-digit numbers? Do the provided activities clearly go beyond face value and expect students to interact with multi-digit numbers at a deeper base-ten understanding?
4. What modifications might be needed to ensure students consistently engage in unitizing activities focused on developing base-ten understanding?

5
Visual Models to Support Additive Reasoning

Big Ideas

- Visual models are a powerful tool for developing conceptual understanding and procedural fluency with additive reasoning.
- Visual models provide an instructional bridge to move students from concrete counting strategies to more abstract number, addition, and subtraction strategies that are based on a strong foundation of number sense.
- An important consideration is to make sure the visual models used for instruction are explicitly connected to each other and support the mathematics being learned.
- Students should be able to interact with a variety of given visual models, sketch models to solve problems, and select a model that makes sense for the situation.
- Number lines are an important visual model that can be used throughout elementary and secondary school years to help students make sense of mathematical relationships and operations.

This chapter focuses on the importance of visual models in the development of number and additive reasoning, as well as the importance of intentionally linking models both to each other and to important mathematical ideas. The focus is on the potential for visual models to support the development of procedural fluency in additive reasoning with understanding. According to the National Research Council, procedural fluency "refers to the knowledge of procedures, knowledge of when and how to use them appropriately, and skill in performing them flexibly, accurately and efficiently" (Kilpatrick et al., 2001, p. 121). Researchers indicate that fluency and conceptual understanding should be built using visual models to develop mathematical ideas rather than memorizing rules and procedures. Several studies have shown that the use of concrete and visual models in mathematics instruction improves student understanding and achievement, across all ability and grade levels (Clements, 2000; Clements & Sarama, 2014; Sowell, 1989).

In this chapter we use the term visual model to refer to both physical and representational models that enhance student thinking related to mathematical concepts. Both are tools that allow students to make sense of, discuss, reflect on, and solve mathematical tasks. Students should be able to both interact with given models and create their own models as they work towards developing procedural fluency with additive reasoning. We will consider the benefits of various models as well as the significance of linking these models explicitly to each other and to the mathematics being developed.

 Chapter 3 The Development of Counting and Early Number Concepts and Chapter 4 Unitizing, Number Composition, and Base-Ten Understanding for more detail on research-based strategies for learning and teaching the mathematics discussed in this chapter.

The importance of number sense for student learning of mathematics was discussed in Chapter 1. While it is often difficult to identify specific features of number sense, Kalchman, Moss, and Case (2001) propose that one indicator is the ability to move among different representations and to use the most appropriate representation in solving a problem. This further supports the argument for providing instruction on the use and structure of visual models throughout the process of developing mathematical proficiency. Consider the vignette below as you begin to think about the role and importance of visual models in mathematics classrooms for our youngest learners.

Importance of Visual Models

Mrs. Brown, a second-grade teacher, gave her students the addition task below at the end of a math lesson.

> 46 children were at the zoo.
> Then 35 more children came to the zoo.
> How many children are at the zoo now?
> Show or explain how you know.

While examining the student work from her classroom she found that some students were counting all or counting on from one number by ones to find the answer. Mrs. Brown wondered how she could help students move away from counting strategies to more sophisticated strategies based on number relationships and understanding. She knew that as the numbers get larger, counting strategies they were using would not be efficient or accurate, but she was at a loss for how to move her students forward towards strategies based on additive reasoning.

This scenario may sound all too familiar to most classroom teachers. The *OGAP Addition Progression* can provide instructional guidance to teachers in helping students move from counting strategies to additive strategies. Specifically, strategies in the *Transitional* level can provide an important bridge to numerical reasoning and fluency. In fact, Mrs. Brown used the progression to consider next instructional steps and observed that *Early Transitional* strategies such as the use of base-ten blocks or number lines could help to move her learners away from counting by ones while still allowing them to use strategies that made sense to them.

When examining all the progressions within the *OGAP Additive Reasoning Framework,* note the role that visual models play in transitioning students to more efficient methods for solving problems. This reflects research-based findings on the effectiveness of visual models for improved student understanding, mathematical abilities, problem solving and reasoning, and the importance of using multiple representations to develop understanding of mathematical concepts. (Gersten et al., 2009; Pape & Tchoshanov, 2001). Visual models play a central role in developing procedural fluency with conceptual understanding, but special consideration should be given to which visual models support and extend student understanding of the key concepts. In the vignette above, focusing on conceptual understanding along with the use of visual models that support grouping or unitizing could substantially improve access to the mathematical goals of the lesson, providing greater access for more students.

While visual models are essential for developing conceptual understanding and flexibility, it is important to remember that they are a means to the mathematics; in other words, they provide access to otherwise complex mathematical ideas but are not the mathematics themselves. Many researchers make the point that manipulatives and models "do not 'carry' the mathematical ideas" (Clements & Sarama, 2014, p. 278), and that the relationships between manipulatives and the concepts they are meant to enhance are not always obvious to students. Although visual models can provide access to mathematical ideas, the ultimate goal is for students to connect the model to more abstract ideas, moving beyond the model. Not all students will be prepared to make the transition away from a visual model at the same time. The role of a teacher involves knowing which models are most effective for making a math concept accessible to students, understanding the relationships between models, and knowing which model(s) will help a student feel most successful. When teachers help students connect and move between visual models to make sense of mathematical situations, the models become more flexible and usable tools. In order to accomplish these things, it is essential that teachers understand the features of different visual models as well as their strengths and weaknesses.

The depth of understanding of a mathematical idea is related to the strength of connections among mathematical representations that students use and have internalized (Pape & Tchoshanov, 2001). The visual model a student uses can often be an indication of their developing understanding, and also an important clue for how instruction can build and extend that understanding to the use of more sophisticated and efficient models.

Take a minute to examine the progressions which make up the *OGAP Additive Framework.*

- Which visual models are prevalent in the progressions?
- Which visual models are unfamiliar to you?
- Are there visual models you anticipated seeing that are absent?

You likely noticed that visual models play a significant role in the learning progressions and that some models are more prevalent than others. Throughout the rest of this chapter we examine the potential of different visual models to support various aspects of additive reasoning. We highlight the visual models that have proven effective in providing access to additive reasoning and procedural fluency for all students at different levels of the progression.

Models that Support Early Number and Unitizing

In this section we examine visual models that support students in anchoring numbers to 5 and to 10. One of the goals of these models is to aid in the development of unitizing towards base-ten fluency.

 Go To Chapter 3 and Chapter 4 for more on the importance of unitizing in understanding number and base ten.

Five- and Ten-Frames

Five- and ten-frames are a powerful visual model for helping students make sense of many of the early number concepts discussed in Chapter 3: Cardinality, part-whole, relative magnitude, hierarchical inclusion, addition, and subtraction. They also frame the quantities 5 and 10, providing a visual model that supports anchoring to the first benchmarks for young learners. At first students may need to count the counters or dots in the frame to determine the quantity, but ultimately the goal would be for students to recognize, compare, and manipulate quantities visually. How students interact with the models gives a teacher insight into a student's number concept development and valuable instructional information. Consider the model in Figure 5.1 that illustrates the quantity 3.

Figure 5.1 A five-frame illustrating 3

 Some questions you might ask a student about this five-frame include:

- How many do you see?
- What goes with the 3 to make a 5?
- How far is 3 from 5?
- Which is greater, 3 or 5?
- Is three closer to 0 or 5?
- What would you take from 5 to get 3?
- Show one more than 3. What number is it?

These questions focus on developing early number concepts as a visual exercise rather than a procedure. The ability to see the quantity without counting is referred to as *subitizing* and five-frames can be a powerful model for subitizing as well as developing concepts such as one more than, one less than, part-whole, relative magnitude, and the inverse relationship between addition and subtraction with quantities up to five.

 Go To Chapter 3 The Development of Counting and Early Number Concepts for more about subitizing and early number concepts.

One variation on five-frames is a bead model (5-bead stick) shown in Figure 5.2 and another is "math hands." Math hands, as mentioned in Chapter 3, can be used to develop many early number concepts. Students can be asked to "show 4 on your hand" and then asked to "show the number that is one less than 4" or "tell what number goes with 4 to make 5." Since all three of these visual models focus on 5 as the anchor number, they can also be easily and explicitly connected to each other. The predictability of how 5 is framed and consistently represented in each of these models benefits students who may be struggling. All these models represent 5 as a unit that is made up of 5 ones, and therefore can be used to support the development of *Early Unitizing* strategies on the *OGAP Base Ten Number Progression*. The questions suggested in Figure 5.1 could be used with all three of these models.

Figure 5.2 Five-bead stick showing 3

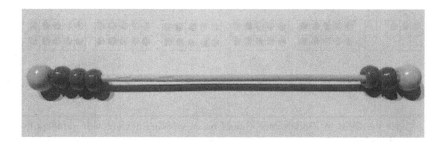

As students develop flexibility with numbers to 5, including an ability to anchor other numbers to 5, the next step is to extend the work to 10. The use of ten-frames to accomplish this work makes sense as a ten-frame is simply an extension of a five-frame. When the time is right, have students find a partner and put their two empty five-frames together to make a ten-frame, taping them on the longer dimension as shown below in Figure 5.3.

Figure 5.3 Two five-frames taped together to form a ten-frame

This kind of explicit instruction around connections between visual models helps students understand the relationship between different models. The concepts previously developed with the five-frames can be easily transferred to a ten-frame. The types of questions suggested for Figure 5.1 can be extended to anchor the remainder of the single-digit numbers to 10 as well as 5.

Ten-frames can also provide a model for skip counting by tens and early base-ten understanding, as they show the ten-to-one relationship. This makes them an important tool for developing this understanding and eventually strategies that involve unitizing by composite units of ten (*Transitional* strategies on the *OGAP Base Ten Number Progression*).

Ask students to count by tens, coordinating the flashing of a full ten-frame with each multiple of ten in the count sequence. When students are orally counting "10 – 20 – 30 – 40 – etc." the teacher is visually representing the number with the ten-frame. When students have counted to 60, pick up the ten-frames and ask, "How many ten-frames does it take to make 60 or how many tens in 60?" This same exercise can be done when counting by tens "off the decade" as well. See Figure 5.4 for an illustration.

Figure 5.4 Counting by tens off the decade using ten-frames

Teacher shows one frame at a time while prompting students to say the total.

Teacher shows:

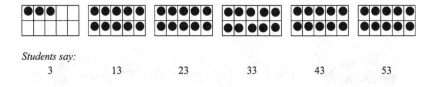

Students say:

| 3 | 13 | 23 | 33 | 43 | 53 |

Since the focus in Kindergarten should be on working flexibly with numbers to ten and understanding teen numbers as "ten ones and some more," five and ten-frames provide powerful visual representations to work towards this goal. For example, illustrating 14 as one full ten-frame and a ten-frame with only 4 dots, as shown in Figure 5.5, allows students to visualize the 14 ones as 10 ones and 4 ones. This is the first step in developing base-ten understanding and will be helpful for students in making the transition to "one ten and four more" in first grade when seeing the same model. The *Base Ten Number Progression* provides an illustration of this transition and provides instructional guidance towards this end.

 Chapter 4 Unitizing, Number Composition, and Base-Ten Understanding for more on these concepts.

Figure 5.5 Fourteen with ten-frames. It can be seen as *ten ones* and four more or *one ten and four more*

Just as with five-frames, there are a number of other models that connect seamlessly to ten-frames including ten-bead sticks and math hands with both hands. Again, the benefit of each of these models is that the discrete items can be counted or moved individually and seen as ten ones, yet simultaneously the unit of ten can be seen as made up of two fives or one ten. Figure 5.6 provides an illustration of each of the models. The structure of all three visual models allows for explicit connections between them, allowing for students to transfer understanding developed with one model to the others. Again, the framed nature of these models provides opportunities for comparing numbers and relative magnitude. Take a minute to examine Figure 5.6 and compare and contrast the three models.

Figure 5.6 Ten-frame, ten-bead stick and math hands

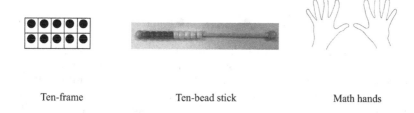

Ten-frame   Ten-bead stick   Math hands

Extending Ten-Frames to Ten-Strips

While ten-frames are a valuable tool for students in early primary grades, they can also be extended to develop understanding of base-ten models that students will continue to use in later grades. We can take the ten-frame back apart into two five-frames and then put it back together vertically as illustrated in Figure 5.7. This should be done with students so they can see the relationships between the models. Note the use of a dark line at five providing a visual benchmark for students.

Figure 5.7 Transforming a ten-frame to a ten-strip

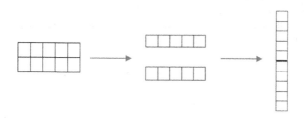

Ten-strips offer some benefits that the ten-frame doesn't offer. We will consider what some of these are by examining Figure 5.8.

Figure 5.8 These ten-strips illustrate the numbers 9 and 8. How could they be used to find the sum of 9 + 8?

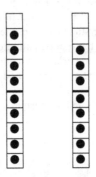

One benefit of the ten-strips set up next to each other is that strategies we teach students for deriving math facts are easily seen. For example, a student can see that physically or visually sliding one from the 8 to the 9 makes a new problem: 10 + 7 which is an easier problem to solve. This is a strategy called "make a ten" in many math programs Since no counters were added to the two strips and 10 + 7 = 17, then 9 + 8 must also equal 17. This is just one of many fact fluency strategies that are easily illustrated using ten-strips.

 Chapter 9 *Developing Math Fact Fluency* for more about the use of ten-strips as a tool for deriving strategies for fluency with addition facts.

Another benefit of transitioning ten-frames to ten-strips is the visual connection to base-ten blocks, which are a prevalent model in most math programs, as well as one of the visual models mentioned in the *K–5 Progression for Number and Operations in Base 10* (Common Core Standards Writing Team, 2019). Figure 5.9 illustrates how the transition between these visual models reflects the transition from early number concepts to base-ten concepts. Instruction that explicitly links these visual models will help ensure a transfer of conceptual understanding from one model to the next as the concepts are developed more deeply.

Figure 5.9 Transition of framed number models to base-ten blocks

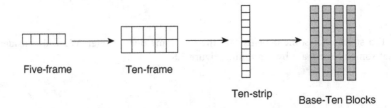

Five-frame   Ten-frame   Ten-strip   Base-Ten Blocks

Chapters 3 and 4 discussed the importance of unitizing, the ability to see a collection of ones as both a group and a set of individual ones. Ten-frames, ten-strips, and base-ten blocks are all models that provide an opportunity to develop unitizing and at the same time allow a student to see the ones that make up one ten. Visual models that support unitizing can move students towards efficiency and deeper understanding of our base-ten number system. In the *OGAP Base Ten Progression*, base-ten blocks appear in the *Transitional* levels of the progression as they are a model that unitizes the ten while still representing the ones.

In Figure 5.10 two students solved a task about packs of ten by drawing visual models to make sense of the problem. Examine Simon and Marshawn's responses to the task and consider how the visual models helped them each solve the problem. How are the two visual models different and where might these solutions fall on the *OGAP Base Ten Progression*?

Figure 5.10 Two student solutions for a base-ten task

Shonda bought packs of bracelets to give out to all the kids in her grade. Each pack has 10 glow bracelets on it. There are 56 kids in Shonda's grade. How many packs of bracelets did she need to buy? Show or explain how you know.

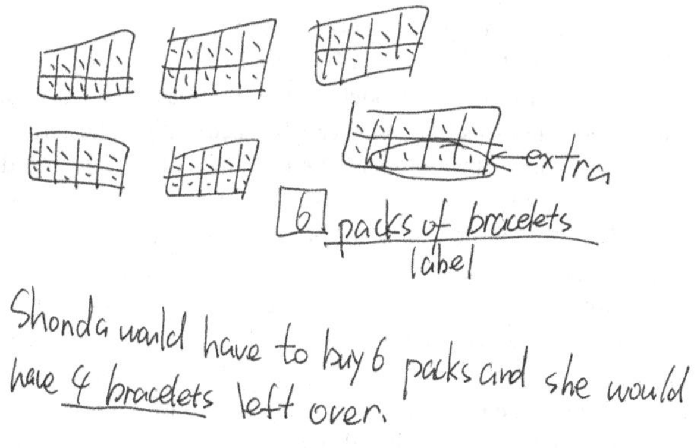

Simon's solution

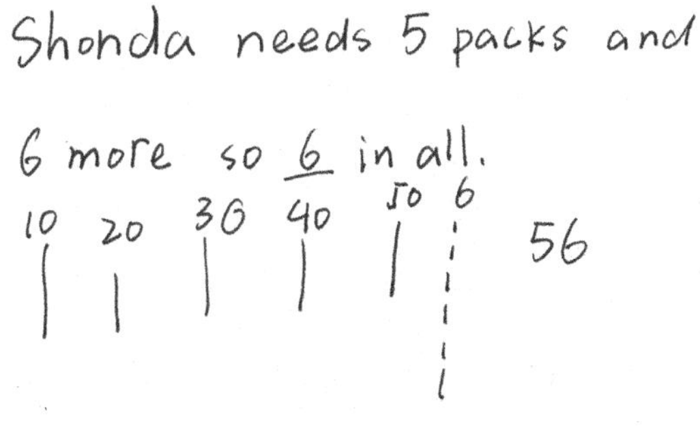

Marshawn's solution

Both Simon and Marshawn solve the task correctly by sketching a visual model. While Simon's visual model is accurate and effective, he illustrates the ten-frames by drawing every dot, each representing a bracelet, resulting in a time-intensive strategy. Simon's solution is an early unitizing strategy on the *OGAP Base-Ten Progression*. Instructionally the next step for Simon may be to move him to a more efficient model, perhaps using Marshawn's solution to make that transition by connecting the two solutions. Marshawn draws base-ten blocks accurately to solve the problem, but represents each ten as one line, a more efficient model. His strategy falls in the next level: *Unitizing by Composite Units*. Simon's model supports his recognition that there are four bracelets left over while Marshawn does not mention the leftovers.

Students will use a visual model they need to understand and solve a task, but not all visual models are efficient for the long term. Instructionally, the role of the teacher is to move students to more efficient models when the time is right. It is likely that both these students chose to use a model that was just right for them and gave them access to the task; however, in time they will need more efficient models and strategies to solve more complex problems with larger quantities.

Ten-Strips to Base-Ten Blocks

Base-ten blocks are a visual model commonly used in math curriculum materials across grade levels. As illustrated on the *OGAP Base Ten Number Progression*, base-ten blocks can be an effective transitional model for building base-ten understanding. Just as with ten-frames and ten-strips, the ones can be seen within the ten and hundred blocks. However, unlike these other models, the markings are subtle and do not indicate any visual benchmark of five within the ten. Numbers between one and ten have to be represented with ones, and tens cannot be taken apart to make ones. Sometimes students recognize that the blocks represent different units that can be counted (ones, tens, hundreds) without really understanding the ten-to-one relationship between the units. For example, a student may draw out a model to represent 463 accurately as in Figure 5.11 and think of it as "4 blocks, 6 sticks, and 3 dots" without any evidence of understanding the actual value of the digits and the relationship between them in 463. This idea of face value versus place value is discussed in greater detail in Chapter 4.

Figure 5.11 Jesse's sketch of 463

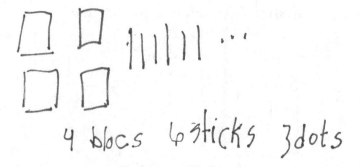

Number Lines as a Tool for Developing Base-Ten Understanding, Addition, and Subtraction

Number lines are powerful and flexible visual models that should play an important role in students' development of base-ten understanding, addition and subtraction. Because of the spatial and linear nature of number lines, they are particularly effective as a model for making sense of the relationships between numbers and developing relative magnitude of number. Relative magnitude, as discussed in Chapter 1, is the understanding of the size of a number in relation to other numbers. Once students have developed understanding of the features and attributes of number lines, they can be effectively employed to develop base-ten understanding, compare and order numbers, and add and subtract flexibly with whole numbers. As students extend their math understanding to more complex numbers and situations in upper grades (e.g., fractions, decimals, negative numbers), number lines continue to play a key role in deepening their understanding.

Number lines are first used in many elementary math program materials for teaching addition and subtraction with an assumption that students already understand the attributes and can use them with understanding; however, that is often not the case. At first number lines can be challenging for students as they require a shift in counting a discrete number of objects to understanding numbers as a unit of length. When using a structured number line, one that has at least two given points that establish an implied unit, a significant challenge for some students is recognizing the relationship between distance and number placement. Examining the variation in responses to a task in Figure 5.12 that asks students to place a number (19) on a structured number line with an implied unit of 5 illustrates some of these challenges and developing understanding.

Figure 5.12 Two student solutions to a number line task

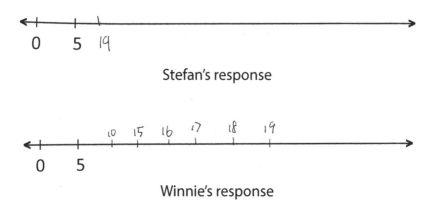

Place 19 on the number line below.

0   5   19

Stefan's response

0   5

Winnie's response

Figure 5.12 shows two student responses with a lack of understanding of the spatial–number relationship in this task. Stefan seems to place 19 on the number line correctly after 5 but without any consideration for the role the distance from 0 to 5 plays in the task. Winnie uses the distance from 0 to 5 to place 10 but then treats every unit after that as equal to the distance from 0 to 5. Both students do not attend to the importance of the implied unit in relation to the location of 19 on the number line. Stefan appears to guess or provide no supporting evidence of reasoning; his strategy would be placed in the *Not Counting or Comparing* section of the OGAP *Base Ten Number Progression*. Winnie has a strategy for counting up to 19 but does not maintain the relative size of the unit so her work would be placed in the *Early Counting* level. Together, these examples illustrate the challenges students face as they make sense of the linear nature of a number line.

In Figure 5.13 both Asha and Beni show that they can place 19 on the structured number line in the correct location. Asha appears to need to see the ones between the fives but reasonably places all marks on the number line. Beni can jump by fives, maintaining the implied unit, and then hops back one to place 19 within reason. Asha's work would be placed in the *Early Unitizing* level since she is able to compose in groups when the ones are visible, while Beni's solution illustrates a strategy in the *Unitizing by Composite Units* level.

Figure 5.13 Two student solutions to a number line task

Place 19 on the number line below.

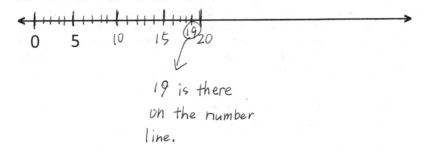

Asha's response

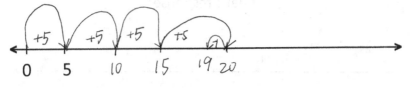

Beni's response

Ten-structured Bead Strings

One way to provide access to understanding of number lines is to include ten-structured bead strings created with 100 beads with groups of 10 beads in alternating colors, as shown in Figure 5.14. This model, developed by Dutch researcher Adri Treffers, allows young students to use their understanding of counting to make sense of the attributes of a number line: as the length from zero increases, the number of beads also increases. This helps establish the relationship between number and length that exists on a number line (Klein, Beishuizen, & Treffers, 1998). Since counting is often a first sense-making strategy for young children, the beads provide the simplest strategy to access the idea of distance on a concrete version of a number line. While counting by ones is not efficient, it is accessible to most students; with explicit instruction connecting the bead string to empty number lines, students can develop understanding of unitizing by tens and more complex number line models.

Figure 5.14 Ten-structured bead string

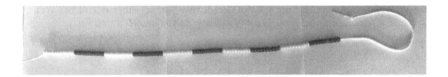

For example, locating the number 19 on the bead string can be accomplished in a few different ways. In Figure 5.15 the quantity 19 is indicated by a small clip placed after the 19th bead. Students often use the ten structure of the bead string to locate 20 and then go back one bead to find 19 but it is also possible to count all 19 beads by ones, or one ten and 9 beads. By providing multiple opportunities to solve problems using this visual and concrete model, teachers can support students to use the tens structure to unitize by tens to locate and compare numbers.

Figure 5.15 Locating 19 on a ten-structured bead string

 Below are some sample questions that could be used with students towards developing fluency of number on the bead string.

- Locate 32 and count by tens. What is the last number you will land on before 100?
- Locate 47. How many tens and ones in 47?
- Locate 64. What are the two decade numbers on either side of 64? Which decade number is 64 closest to and how do you know?
- Locate 43. Locate 59. Which number is closer to 50? Explain how you know.
- Locate 71. What number is ten more than 71?
- Locate 38. What goes with 38 to make 100?

Once students become fluent making sense of numbers on the bead string, instruction can focus on transferring work done on the bead string to a student-generated sketch, as illustrated in Figure 5.16. Helping students to make this connection will allow for a transfer of the concepts developed on the concrete model to the visual sketched model of empty number lines.

Figure 5.16 Locate 64. What are the two decade numbers on either side of 64? Which decade number is 64 closest to and how do you know?

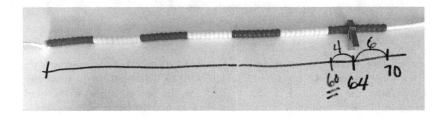

Number Paths

A *number path* like the one shown in Figure 5.17 is another model that can be used to locate and compare numbers and make connections to number lines. It shares many of the attributes of a bead string but the numbers are in a fixed position. Instead of a bead representing a quantity, each number has its own corresponding space. Figure 5.17 is an illustration of a section of a number path where color has been utilized to illustrate the decades and easily connect to the bead string. Students are more likely to locate numbers by finding the numeral on the number path than by focusing on the units of one and ten, but by representing the decades in different colors, students can see the tens and ones in any number as well as how close a number is to the next decade. As with bead strings, the linear aspect of the number path illustrates the concept that the further a number is from zero, the greater the magnitude.

Figure 5.17 One section of a number path that goes from 1 to 100

23	24	25	26	27	28	29	30	31	32	33	34	35	35	37	38	39	40	41	42	43	44	45

Number paths can be connected to bead strings and also to *hundred charts*, a visual model used by many math programs as early as kindergarten for base ten, addition and subtraction concepts and strategies. Having students shade alternating rows of a hundred chart (as illustrated in Figure 5.18), and then cutting the rows apart and taping them back together as a number path (Figure 5.17), can help them understand the various models and their relationships.

Figure 5.18 A hundred chart ready to be turned into a number path

1	2	3	4	5	6	7	8	9	10
11	12	13	14	15	16	17	18	19	20
21	22	23	24	25	26	27	28	29	30
31	32	33	34	35	36	37	38	39	40
41	42	43	44	45	46	47	48	49	50
51	52	53	54	55	56	57	58	59	60
61	62	63	64	65	66	67	68	69	70
71	72	73	74	75	76	77	78	79	80
81	82	83	84	85	86	87	88	89	90
91	92	93	94	95	96	97	98	99	100

Hundred charts have some advantages in highlighting patterns in the number system, but as many teachers can attest, they also have a number of features that make them challenging for students to use. For example, locate 34 on the hundred chart in Figure 5.18. Think about the question, "Is 34 closer to 39 or 44?" Of course, we know that in terms of magnitude 34 is closer to 39 but visually on the hundred chart it is literally closer (in distance) to 44. For another example, think about moving down one space from 52. This move is not adding one to 52 but rather ten, but the ten is not easy to see. The ten is made up of the 8 squares that complete the decade and then the 2 squares all the way back on the left of the next row, as shown in Figure 5.19.

Figure 5.19 62 is ten more than 52 on a hundred chart

1	2	3	4	5	6	7	8	9	10
11	12	13	14	15	16	17	18	19	20
21	22	23	24	25	26	27	28	29	30
31	32	33	34	35	36	37	38	39	40
41	42	43	44	45	46	47	48	49	50
51	52	53	54	55	56	57	58	59	60
61	62	63	64	65	66	67	68	69	70
71	72	73	74	75	76	77	78	79	80
81	82	83	84	85	86	87	88	89	90
91	92	93	94	95	96	97	98	99	100

+ 8

+ 2

When students work with hundred charts, they need to know that moving one space right or left is a move with a value of one but moving one space up or down has a value of ten. In his research on the role of the hundred chart in children's conception of ten, Cobb (1995) found:

> ...children's use of the hundred board did not support their construction of increasingly sophisticated conceptions of ten. ...It appears instead that the children's efficient use of the hundred board was made possible by the construction of increasingly sophisticated place value conceptions.
>
> (pp. 375–6)

In other words, the hundred chart was helpful for students who had already developed the understanding that one ten is equivalent to ten ones but did not support students in developing that understanding. Many students end up memorizing shortcuts for using the hundred chart without truly understanding that they are adding or subtracting a unit of ten or ten ones. For students who struggle with unitizing it can be difficult to distinguish between these units.

Based on this knowledge, teachers may want to decrease the use of hundred charts as a visual model in the early grades in favor of number paths and bead strings. Once understanding of unitizing has been developed, number paths can be used to help make sense of hundred charts for students, and at the same time connect ten-structured bead strings to hundred charts, making it clear to students that all these models illustrate the same concepts in various forms.

Another critical step towards the use of more flexible number line models is to link number paths to a number line where all whole numbers are visible. As can be seen in Figure 5.20, a number line with all numbers visible has been drawn directly under a number path. Observe that the hash marks are placed at the end of the space for the corresponding number on the number path. When introducing and connecting these models it is important to fill in the numbers with students, emphasizing that each hash mark lines up with the end of each block on the number path.

Figure 5.20 Number paths connected to number lines

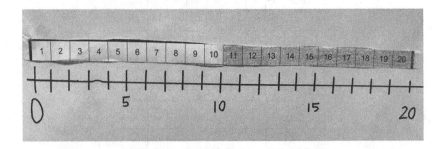

Structured Number Lines

Structured number lines have two or more given points, requiring students to work within the implied unit. As discussed earlier, Figures 5.12 and 5.13 are examples of tasks that have two given points. When students first begin to work with structured number lines, it is typical for them to ignore the implied unit and just place numbers (Figure 5.12). Once they begin to understand the relationship of length to number, students often work with smaller jumps of ones and fives, moving to larger unit jumps as they develop flexibility and confidence with the model. In Figure 5.21, students were asked to place 25 on the number line. The four solutions illustrate increasingly efficient strategies students will use to solve the relative magnitude task. Examine the student responses to the task and consider where each solution would be placed on the *OGAP Base Ten Number Progression*.

Figure 5.21 Four student strategies for placing 25 on a structured number line

Place 25 on the number line below.

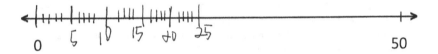

Anne's solution

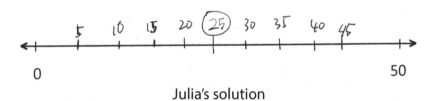

Julia's solution

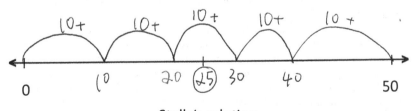

Stella's solution

Figure 5.21 Continued.

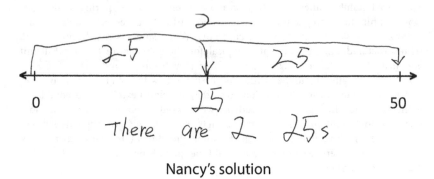

Nancy's solution

Anne's work reflects an *Early Unitizing* strategy because the evidence shows the ability to compose in groups when the ones are visible. Both Julia and Stella's work reflect a *Unitizing by Composite Units* strategy because they locate 25 by composite units of 5 and 10 respectively, without needing to see the ones. Nancy's work shows evidence of *Number Composition by Place Value Parts* as she is able to use number relationships with an implied understanding of place value when considering relative magnitude. Nancy understands that there are two groups of 25 in 50 and therefore 25 is halfway between 0 and 50.

Empty Number Lines

Empty number lines do not have predetermined units; they are simply unmarked lines that allow a student to record numbers and jumps or hops for comparing, adding, and subtracting quantities to solve problems. Since there are no given points on an empty number line, students can focus on communicating their thinking without concerning themselves with the implied unit. The empty number line can be a flexible model for recording student thinking and at the same time a tool for solving a problem visually. Fosnot and Dolk (2001), building on the work of Gravemeijer (1999), propose that as students use visual models "they will move from models *of* thinking to models *for* thinking" (p. 81). This transition happens as they construct models that can help them think more flexibly and strategically about their solutions as opposed to only communicating their thinking after having solved the task.

The benefits and flexibility of the empty number line make it a powerful visual model; math education researchers, as well as math programs, have increasingly recommended its use as a transitional model for students to use as they construct understanding of complex math ideas (Gravemeijer & van Galen, 2003; Verschaffel et al., 2007).While sharing the linear feature with other number line models, empty number lines are unique in the sense that they offer a blank canvas, lacking any detail, allowing students to bring multiple strategies to the task. The openness of the model increases the likelihood that students at varying levels of understanding can access the mathematics and communicate their strategy effectively. This can be observed in Figure 5.22 where Tari and Bo have used an empty number line to illustrate whether 50 is closer to 26 or 91.

In both students' solutions their base-ten understanding is visible, and their ability to use the empty number line to communicate the solution to this relative magnitude question is effective. While Tari uses units of tens to compare the numbers Bo is able to chunk tens to illustrate the distance of 50 to 26 and 91. Because of the open flexible nature of the empty number line both students were able to approach the task at their own level of understanding. Empty number lines can be an effective and flexible tool for comparing numbers as well as for addition and subtraction of multi-digit numbers.

Figure 5.22 Tari and Bo's solutions

Is 50 closer to 26 or 91? Explain how you know.

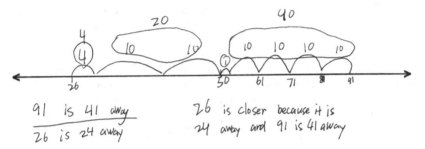

Tari's solution

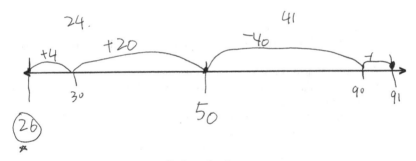

Bo's solution

When first introducing students to empty number lines, teachers can use them during whole class discussions by sketching them under more concrete number lines such as structured bead strings or number paths, as shown in Figure 5.23, to illustrate students' verbal explanations of strategies for solving a task. Later students can employ the empty number line to communicate their own thinking independently and as a tool for solving a variety of number-related tasks.

Figure 5.23 A teacher connecting a number path to an empty number line

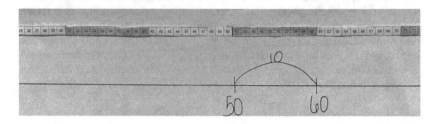

 Chapters 6 and 7 on Addition and Subtraction for more on the use of empty number lines for addition and subtraction of multi-digit numbers.

As seen throughout this chapter, number lines can take many forms, and these forms vary in the degree to which they are concrete and accessible. Teachers can strategically select and use number line models to help students understand the format and attributes associated with each, towards the goal of flexibly understanding and using empty and structured number lines. The *OGAP Number Line Continuum*, which can be downloaded at **www.routledge.com/9780367462888**, is presented as an instructional tool that provides guidance for teachers to offer multiple access points for learners at each grade level based on their developing understanding. This tool provides an instructional path for all children to develop deep and flexible understanding of the number line. When examining the *OGAP Number Line Continuum* consider:

- the variation from grade to grade,
- how different visual number line models can be utilized to make sense of numbers,
- the connections that can be made between models,
- how the continuum can be used to support students who are struggling with number lines, and
- the overall path towards the goal of fluency with empty number lines in grade 2 for addition and subtraction.

CCSSM and Visual Models

The CCSSM calls for students to use visual models (concrete and pictorial) to solve problems and make sense of concepts. These models are intended to be used as tools and strategies to support students' development of number, base ten, and addition and subtraction in grades K–2. In addition, one of the Standards for Mathematical Practice, "Use Appropriate Tools Strategically," asks students to consider available tools when solving a math task, including concrete models.

Throughout the K–2 standards, the CCSSM explicitly states that students should use objects, concrete models, or drawings to solve problems involving number, relative magnitude, addition, and subtraction. In the *Progressions for the Common Core State Standards for Mathematics*, the authors specify various models that should be used by students and teachers to help develop understanding of the important concepts in each grade level.

- In Kindergarten, children use counters, framed models such as five- and ten-frames, number paths, and number bonds, to develop understanding of number.
- In grade 1, students use concrete ten-structured models such as base-ten blocks and bead strings to develop understanding of base-ten concepts.
- In grade 2, children use number lines to compare and order numbers and add and subtract multi-digit numbers.
- In every grade there is a focus on students' use of drawings and pictorial representations of the concrete models to transfer to more efficient strategies and develop fluency.

In grade 3 and beyond the expectation is that students will build procedural fluency with operations with larger numbers based on a strong foundation with visual models. As students develop fluency with two-and three-digit numbers, the goal is to be able to use written methods, "though objects and drawings can be used with explanations to overcome errors and to continue to build understanding as needed" (Common Core Standards Writing Team, 2019, p. 62).

Additional Considerations When Using Concrete and Visual Models

While we have not considered all the possible visual models that can be used to help young learners make sense of complex math ideas, we have highlighted some of the more flexible and effective models. As a teacher, when incorporating visual models in your classroom instruction consider the following:

- Models should enhance the mathematics, and not compromise the concepts being learned. Be aware of the strengths and weaknesses of various models when deciding which visual models to use with students.
- Students should be able to build, sketch, and interact with models to make sense of the important mathematics. Beware of using visual models or manipulatives that are overly structured and inflexible (such as base-ten blocks). When students participate in constructing the models or composing and decomposing them, they perceive them as more flexible and can see themselves as involved in developing the mathematical ideas.
- Use a few effective visual models well. Consider teaching multiple concepts with a few manipulatives or models versus using a new model with every new concept. According to Hiebert and Wearne (1996), deeper experience with one manipulative is more effective than using many manipulatives for one concept. A few well-structured visual models that are used to make sense of multiple concepts will help prevent students from relating a concept to one specific manipulative and help them understand that models are tools for learning.
- Move from concrete manipulatives to pictorial models that students can sketch and use to make sense of a task efficiently. In this way they can move towards fluency in base ten, addition and subtraction, moving up the progressions to more efficient strategies for solving problems. While initial instruction may include the use of a concrete model to make sense

of the mathematics, as students develop fluency, they should be able to sketch their own visual models.

Chapter Summary

- Visual models play a key role in the teaching and learning of number, base ten, addition, and subtraction. Visual models can be used to develop a strong conceptual foundation and provide students access to mathematics that might otherwise be unavailable.
- Students should be able to use a variety of visual models and understand the relationships between the models.
- Number lines are an important model throughout a student's mathematical learning; the understanding of number lines can be introduced concretely in the earliest grades and progressively connected to more abstract ideas of distance, measurement, and magnitude and with various types of numbers for years to come.

Looking Back

1. **Using Visual Models in Instruction:** Visual models can provide access to mathematical ideas that may be otherwise difficult to understand. When using models instructionally, the goal is for students to connect the model to more abstract ideas. Consider this task:

   What number is 30 more than 43?

   (a) What skills and concepts are needed to find the number on the following visual models?

   - Bead string
   - Hundred chart
   - Structured number line (with at least two given points)
   - Empty number line

   (b) What are the benefits of one visual model over the other?
   (c) What potential misconceptions should teachers be aware of when making instructional decisions about the use of hundred charts for a task like this one?

2. **Models that Support Early Number and Base-ten Development:** Figure 5.24 shows three ten-structured models that help students build understanding of unitizing towards base-ten fluency. In addition, these models illustrate the first benchmarks for young math learners, numbers 5 and 10.

   (a) Compare the features of the three tens-structured models in Figure 5.24.
   (b) What visual features distinguish or highlight the ones, fives, and ten in each model?
   (c) What connections can be made between the models that call attention to the benchmarks of 5 and 10?
   (d) How can these models help build understanding of cardinality, relative magnitude, hierarchical inclusion and the relationship between addition and subtraction?

Figure 5.24 Three ten-structured visual models

Ten-frame   Ten-bead stick   Math Hands

3. **Number Lines for Developing Base-Ten Understanding:** Number lines are powerful and flexible visual models that should play an important instructional role in student learning. Refer to the *OGAP Number Line Continuum*, which can be downloaded at **www.routledge.com/ 9780367462888**. Consider the features of ten-structured bead strings, number paths, structured number lines, and empty number lines. Strategic instruction that includes these models increases the likelihood that students will understand the attributes of empty number lines before they are expected to use them flexibly to solve problems.
   (a) How do the features of each model visually show relative magnitude, the size of a number in relation to other numbers?
   (b) Look across the models on the continuum. What visual features of the models change from kindergarten to second grade and what features stay the same?
   (c) What features make the empty number line a flexible tool for solving problems?

4. **Making Connections Between Models:** Students should be able to use a variety of visual models and understand the relationships between models. Consider the *OGAP Base Ten Progression* and revisit the important features, concepts, and number relationships discussed in this chapter that must be understood as students move from *Counting* strategies to *Base Ten Understanding*. With a colleague, generate questions that can be used to help students see relationships between the visual models found on the progression.

Instructional Link

Use the following questions to analyze ways your instruction and math program provide opportunities to build fluency with and understanding of important grade-level additive concepts and skills.

1. What visual models are used in your instruction or presented in the math curriculum?
2. Does the math program support intentional use of visual models in ways that are targeted to important additive concepts?

3. When are number lines introduced? Upon introduction, is there attention given to building understanding of number lines and making connections to previously used models?
4. What opportunities does your math curriculum provide students to see relationships between models?
5. How could you build more opportunities for meaningful use of visual models into your instruction?

6

Developing Whole Number Addition

Big Ideas

- The strategies students use to solve addition problems reflect their developing understanding of base-ten number concepts, properties, and relationships.
- There are three main types of additive strategies that involve decomposition and recomposition by place value, incrementing by place value, and changing one or both addends and compensating.
- Strategies for addition should become efficient, accurate, and generalizable over time.
- Standard algorithms offer efficient recording methods for adding large or multiple quantities; transparent algorithms are beneficial in maintaining the place value of the quantities.
- Teaching standard algorithms before solid base-ten understanding is developed can be detrimental in the long run for students' mathematical understanding, computational fluency, and disposition towards mathematics.

This chapter focuses on the development of strategies and algorithms for solving problems involving addition and highlights the importance of visual models, transitional strategies, and transparent and standard algorithms in bridging from counting to additive strategies.

The *OGAP Addition Progression* illustrates that students' strategies for addition will transition from strategies based on their understanding of counting to more efficient strategies that are based on their developing understanding of base ten and the properties of addition. The *Additive* level of the progression includes both *strategies* that are developed by students as well as standard and nonstandard *algorithms* learned through instruction. Both strategies and algorithms can be efficient, accurate, and generalizable—the important criteria for computational fluency (Bass, 2003; Russell, 2000).

In the National Research Council's report, *Adding It Up: Helping Children Learn Mathematics*, mathematical proficiency is defined as the intertwining of five components: conceptual understanding, procedural fluency, strategic competence, adaptive reasoning, and productive disposition (Kilpatrick, et al., 2001, p. 116). This chapter illustrates how the five components of mathematical proficiency can be achieved in relation to addition. Chapter 7 is a related chapter that focuses on subtraction.

Early Strategies for Addition

Young children first come to understand addition as accumulating more or joining two collections, and this allows them to solve addition word problems by counting. At first, this is a *counting all* procedure: counting out the first addend, then counting out the second addend, and then counting the entire collection beginning at one. As they develop their understanding of cardinality and hierarchical inclusion, they can make this counting process more efficient by starting with one addend and *counting on* the second one. This shift represents the transition from *Early Counting* to *Counting* strategies on the *OGAP Addition Progression*.

 Chapter 3 The Development of Counting and Early Number Concepts for more on cardinality, hierarchical inclusion, and counting on.

In Figure 6.1, Carla uses an *Early Counting* strategy to solve a problem involving the addition of two parts of a collection. She draws out 9 fingers to represent the red crayons, draws another 15 fingers to represent the blue crayons, and then counts all the fingers up by ones, beginning at one. In Figure 6.2, Ray uses a more efficient counting strategy; he begins with 15 and then counts on 9 ones, using the number line to keep track. His number line illustrates how this strategy involves keeping track of two chains of numbers: 15 to 24 and 1 to 9.

Figure 6.1 Carla's response. Carla uses an *Early Counting* strategy for addition

Coral and Sean are counting crayons. Coral counts 15 red crayons. Sean counts 9 blue crayons. How many crayons did they count together?

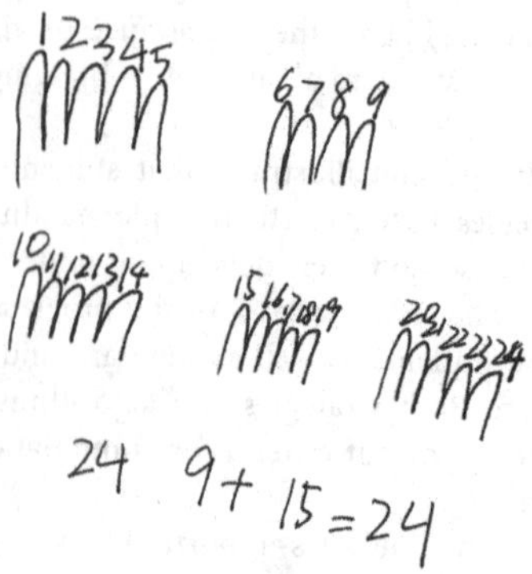

Figure 6.2 Ray's response. The evidence in Ray's work shows a *Counting* strategy for addition

Coral and Sean are counting crayons. Coral counts 15 red crayons. Sean counts 9 blue crayons. How many crayons did they count together?

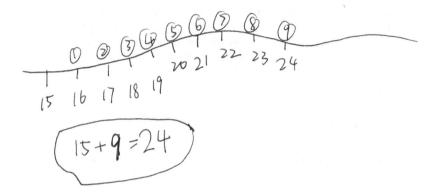

Another important milestone occurs when children begin to develop the ability to *unitize* and recognize that it is easier and more efficient to work with groups or parts when combining quantities. Models such as fingers, number lines, ten-frames, and bead strings can help support this unitizing, a defining characteristic of *Transitional* strategies on the *OGAP Addition Progression*. For example, Carla's drawing of fingers (Figure 6.1) could support unitizing by fives to count 5, 10, 15, 20 and then 4 more. In Figure 6.3, Fatima uses the five structure in the ten-frames to determine the total of 13. She adds the two fives to get 10, and then knows that 10 and 3 more make 13. These visual models can provide an instructional bridge from *Counting* to *Transitional* strategies.

Figure 6.3 Fatima's use of the fives in the ten-frame model illustrates a *Transitional* strategy

Which number is shown? Show or explain how you know.

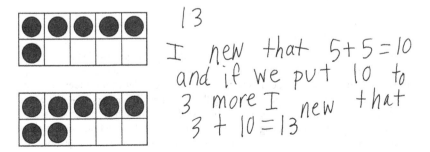

Chapter 5 Visual Models that Support Additive Reasoning for more on the importance of visual models.

With the development of part–whole knowledge and hierarchical inclusion comes the ability to conceptualize all three quantities mentally, which then allows children to remember number combinations and mentally adjust the addends. With this understanding, along with growing knowledge of number combinations (such as doubles) and properties of addition, children can develop *Additive* strategies such as the one shown in Figure 6.4. Kayla adds 9 and 15 by using part–whole understanding to break the 9 up into 5 and 4 and then adds 5 to 15 (a known fact) to get 20, then another 4 to get 24.

Figure 6.4 Kayla uses an *Additive* strategy illustrating part-whole understanding and knowledge of number combinations

Coral and Sean are counting crayons. Coral counts 15 red crayons. Sean counts 9 blue crayons. How many crayons did they count together?

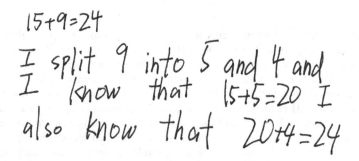

The development of strategies at the *Additive* level is also influenced by the size of the addends and the problem type.

 Chapter 8 Additive Situations and Problem Solving for more on the influence of number size, additive problem types, and structures; Chapter 9 Basic Fact Fluency for more on derived facts for addition.

Properties of Addition

Strategies at the *Additive* level of the *OGAP Addition Progression* depend upon a strong understanding of number as well as understanding of some important properties of addition. Because addition is *commutative*, the order of the addends can be changed or "commuted" without changing the sum; for all numbers, $a + b = b + a$, as illustrated in Figure 6.5. This understanding allows children to make some addition problems easier by starting with the larger amount. Subtraction, on the other hand, is not commutative; if a and b are different numbers then $a - b$ does not equal $b - a$.

In Figure 6.6, the evidence in Sam's solution shows he understands that he can start with 14 and count on 5 even though the problem lists the addends in the opposite order. This is much more efficient than counting on 14 from 5.

Figure 6.5 The commutative property of addition

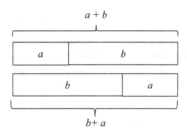

Figure 6.6 Sam's Response. Sam's strategy shows understanding of the commutative property of addition

If Daniel had 5 cupcakes, and he made 14 more, how many cupcakes would he have?

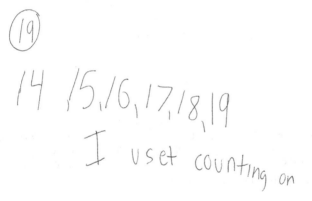

Addition is also *associative*, which means that with more than two addends, the order in which the addition operation is performed does not matter (see Figure 6.7). In the expression *(a + b) + c*, the parentheses indicate that the operation *a + b* should be performed first. However, this expression can be changed to *a + (b+ c)*, where the operation *b + c* is performed first, without changing the result.

Figure 6.7 The associative property of addition

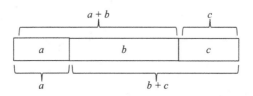

An application of this property can be seen in Figure 6.8 where the problem involves combining three addends (6, 3, and 4). Instead of adding the quantities in the order in which they are presented in the problem, George first adds 6 and 4 to make 10 and then adds 10 and 4, making it into an easier problem to solve. Together the commutative and associative properties allow for quantities to be decomposed and rearranged in many ways to construct efficient and flexible addition strategies.

Figure 6.8 George's strategy illustrates the commutative and associative properties of addition

Shawn started to build a building with 6 blocks. Armondo added 3 blocks onto the building. Then Shawn put 4 more blocks on top. How many blocks did Shawn and Armando use for their building?

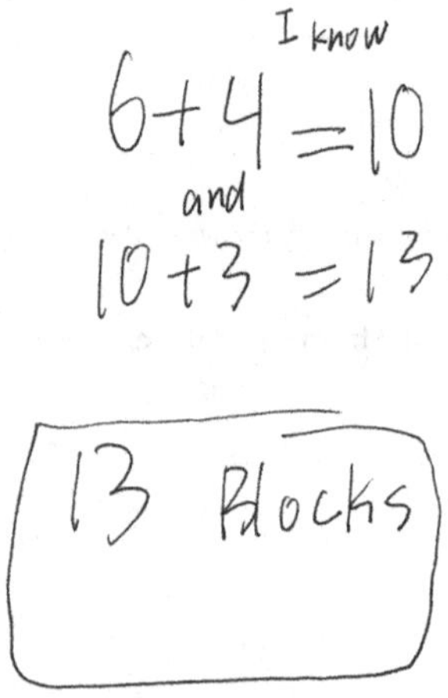

I know

$$6 + 4 = 10$$

and

$$10 + 3 = 13$$

13 Blocks

A third concept that is important for the development of additive strategies is the inverse relationship between addition and subtraction. This understanding can be seen in Fiona's solution in Figure 6.9. To add 15 and 9, she changes the problem into 15 + 10. She then recognizes that she needs to compensate for this change by subtracting one from the total.

Figure 6.9 Fiona's additive strategy shows understanding of addition and the inverse relationship with subtraction

Coral and Sean are counting crayons. Coral counted 15 blue crayons. Sean counted 9 red crayons. How many crayons did they count together?

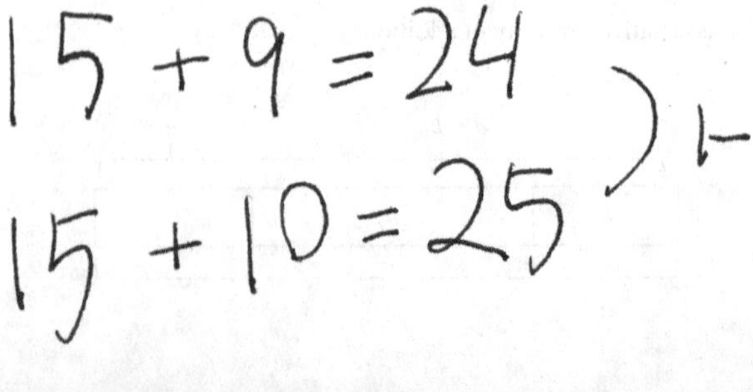

$$15 + 9 = 24$$
$$15 + 10 = 25$$

A related concept is *compensation*—the understanding that a change in one addend can be compensated by an opposite change of the same magnitude in the other addend, with no resulting change to the total or sum. For example, 3 + 4 can be transformed into 2 + 5 by taking 1 away from the 3 and adding it to the 4. The result is an equivalent quantity, as the two colors of blocks and ten frames in Figures 6.10 and 6.11 illustrate.

Figure 6.10 Compensation with cubes of two colors

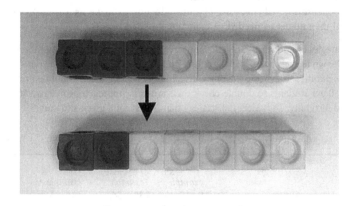

Figure 6.11 Compensation by moving a dot on a ten frame

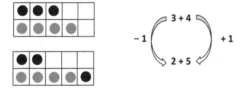

This strategy can be used to calculate 6 + 8, for example, by changing into a known doubles fact, 7+ 7. Solving problems that involve finding all the ways to make a total of 10 (or another sum) is a context that can help students discover these patterns.

Additive Strategies for Multi-Digit Numbers

As children begin to unitize by tens, they can use this developing understanding of base ten along with properties of addition to add multi-digit quantities with increasing efficiency.

There is a wealth of research to show that students can develop their own strategies for efficiently adding multi-digit numbers without being directly taught procedures (Carpenter & Fennema, 1992; Carraher, Carraher, & Schliemann, 1987; Cobb & Wheatley, 1988; Fuson et al., 1997; Hiebert & Wearne, 1996; Kamii, 1985, 1989; Nunes, 1992; Saxe, 1988). Moreover, these strategies are often efficient and based on students' developing understanding of number and number relationships. Many children and adults use these methods for mental calculations because it is easier to keep track of the steps and the quantities. While many variations are possible, these strategies fall into three main categories (Fuson et al., 1997):

1. *Decomposing and Recomposing by Place Value.* Break up the addends by place value parts, add the place value parts or units separately, and then add the subtotals.
2. *Incrementing.* Keep one addend whole and then count or add on the other addend in place value parts.
3. *Compensating.* Change one or both addends to make an easier problem and then adjust as needed to maintain the total.

These strategies are powerful and efficient ways to solve problems with numbers of any magnitude; while each has advantages, it is important that students are exposed to and have opportunities to develop fluency with all three. Having more than one method for addition allows students to double-check their answers, approach problems with confidence and flexibility, and select the strategy that is best for a given situation.

Table 6.1 Additive strategies for one-, two-, and three-digit addition

Strategy	Description	One-digit example 9 + 7	Two-digit example 39 + 56	Three-digit example 135 + 578
Decomposing and recomposing by place value	Break up the number by place value parts, add the place value parts or units separately, and then add the subtotals	N/A	30 + 50 = 80 9 + 6 = 15 80 + 15 = 95	100 + 500 = 600 30 + 70 = 100 5 + 8 = 13 600 + 100 + 13 = 713
Incrementing by place value	Keep one addend whole and then count or add on the other addend in place value parts	9 + 1 = 10 10 + 6 = 16	39 + 50 = 89 89 + 6 = 95	135 + 500 = 635 635 + 70 = 705 705 + 8 = 713
Mixed strategy	Add the larger place value parts and then count or add on the smaller place value parts.	N/A	30 + 50 = 80 80 + 9 = 89 89 + 6 = 95	100 + 500 = 600 600 + 30 + 70 = 700 700 + 5 = 705 705 + 8 = 713
Compensating	Changing one or both addends to make an easier problem and then adjusting to maintain the total.	9 − 1 = 8 7 + 1 = 8 8 + 8 = 16	40 + 56 = 96 96 − 1 = 95	578 + 22 = 600 135 − 22 = 113 600 + 113 = 713

Strategies that reflect a mixture of these approaches are also possible. For example, many students will use a mixture of decomposing and incrementing, indicated in Table 6.1 as a mixed strategy, by first decomposing and adding the larger place value parts and then incrementing by the smaller parts.

In the following sections, each of these strategy types will be discussed separately and illustrated with examples of student work. We also explore how each strategy can be supported with visual models or instructional strategies. Although many children will invent or develop these strategies without being explicitly taught, intentionally supporting this development with visual models is important for ensuring that all students have an opportunity to access and make sense of these efficient and generalizable strategies with understanding.

Decomposing and Recomposing Strategies

In Figure 6.12, Olivia solves a multi-digit addition problem by breaking both quantities into tens and ones. Her work shows the base-ten understanding that 25 is composed of 20 and 5 and 28 is composed of 20 and 8, and that these parts can be added in any order (the commutative and associative properties). Olivia brings the tens together to add them first (20 + 20), then adds the ones, and then adds the two subtotals together.

Figure 6.12 Olivia decomposes both addends by tens and ones and adds them separately

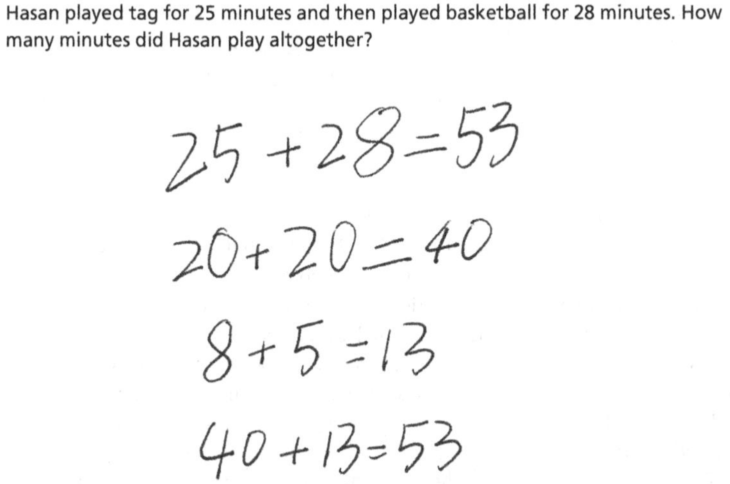

Hasan played tag for 25 minutes and then played basketball for 28 minutes. How many minutes did Hasan play altogether?

$$25 + 28 = 53$$

$$20 + 20 = 40$$

$$8 + 5 = 13$$

$$40 + 13 = 53$$

When students first develop this strategy, they may need to use extra written steps to help them keep track of the decomposition and recomposition. These extra steps are also helpful in supporting them to explain their reasoning. Many teachers introduce a tree or branching representation like the one shown in Figure 6.13 to help students record the decomposition of numbers into tens and ones. As they become more comfortable with the strategy, students can combine steps, moving towards efficiency and standard recording methods and algorithms. Initially developing these strategies may take more time and steps for students to record their thinking, but by using methods that make sense to them, students will be working towards procedural fluency while reinforcing the development of base-ten understanding.

In Brandon's solution, shown in Figure 6.13, you can see how his use of branching (sometimes called *number bonds*) shows the decomposition into tens and ones and also helps him keep track of the recomposition of the tens and the ones into two sums (150 and 14).

Figure 6.13 Brandon's solution shows both the decomposition and recomposition

A movie theater was selling tickets for a new movie. 98 tickets were sold, and now the theater has 66 tickets left. How many tickets were there to start with?

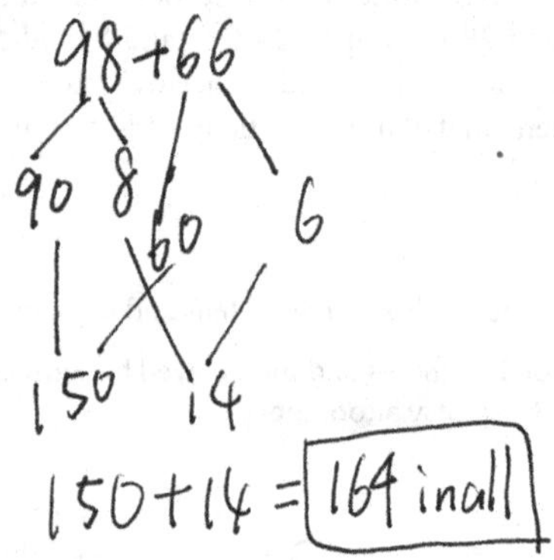

Students will generally work left to right when adding, dealing with the largest quantities first (Fuson et al., 1997; Kamii, 1989), but this strategy can also work right to left. Working left to right is similar to how children read and generates an estimate or reasonable approximation of the answer in the first step, while working from right to left mirrors the Traditional US algorithm for addition, discussed later.

Strategies based on decomposition by place value can also be supported with concrete and visual base-ten models, such as base-ten blocks or ten-frames (Fuson & Briars, 1990). In the example shown inFigure 6.14, Ava first models both numbers with written representations of base-ten blocks. She then groups 10 ones into a 10 and adds the 5 tens and then the ones. Here the representation of the base-ten model helps to distinguish tens and ones so that they can be combined separately. Ava's solution shows that she understands that 10 is made up of ten ones and the ones can be grouped into another 10. This kind of representation of tens and ones units can also help support students in explaining the steps of the strategy (Common Core Standards Writing Team, 2019).

Figure 6.14 Ava represents base-ten blocks to decompose and recompose the addends

Hasan played tag for 25 minutes and then played basketball for 28 minutes. How many minutes did Hasan play altogether?

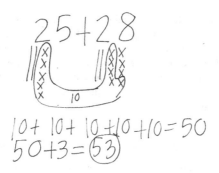

Note, however, that the use of the base-ten block model results in a strategy that is essentially a *counting all* strategy with units of tens and ones. After modeling both numbers in tens and ones, Ava counts all the tens, starting from zero, and then adds on the ones. In this case the model may be supporting a counting-all-by-units rather than an additive strategy. On the *OGAP Addition Progression*, this would be considered an *Early Transitional* Strategy since the student is counting by tens rather than multiples of ten.

 Chapter 5 for more on the limitations of base-ten blocks as a visual model

Incrementing Strategies

The distinguishing factor of incrementing strategies is that one addend is kept whole or intact and the other one is decomposed. In Figure 6.15, to add 98 and 66 Sonya starts with 98 but breaks up 66 into 64 and 2, and then adds the 2 on to 98 to get 100. This then makes it easy to add on 64 to find the total of 164.

Figure 6.15 Sonya uses an incrementing strategy

A movie theater was selling tickets for a new movie. 98 tickets were sold, and now the theater has 66 tickets left. How many tickets were there to start with?

$$98 + 66 = \boxed{164 \text{ tickets}}$$
$$\overset{\diagdown}{64} \quad \overset{\diagup}{2}$$
$$98 + 2 + 64 = \boxed{164 \text{ tickets}}$$

A challenging aspect of this strategy is decomposing one addend into useful parts and then keeping track of the parts. The two solutions in Figure 6.16 illustrate how this strategy can be supported and represented by the empty number line. Yasmin's solution reflects the same strategy shown in Sonya's solution in Figure 6.15, but her thinking is supported by the empty number line. Liam also uses an empty number line but starts with the 66, adds 4 to get the next ten, then adds the remaining 94 (from 98–4) in tens and ones (90 and 4). The 98 that was decomposed and added on in parts is shown in separate jumps. These strategies are considered *Transitional* on the *OGAP Addition Progression* because of the use of visual models. As their understanding deepens, students will be able to use these strategies without the visual support, moving the visual model from one drawn on paper to a mental model.

Figure 6.16 Using empty number lines to support incrementing strategies

A movie theater was selling tickets for a new movie. 98 tickets were sold, and now the theater has 66 tickets left. How many tickets were there to start with?

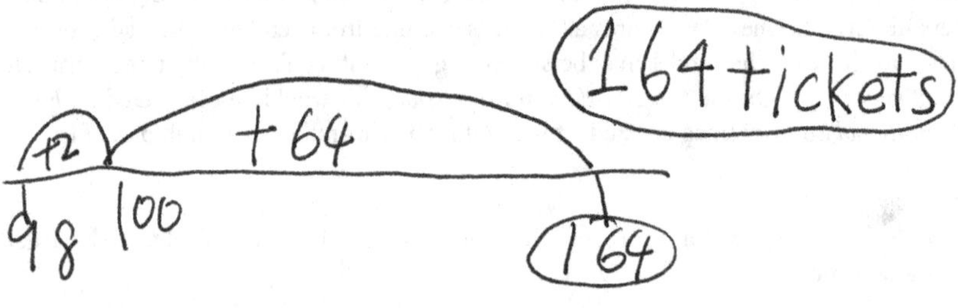

Yasmin's solution

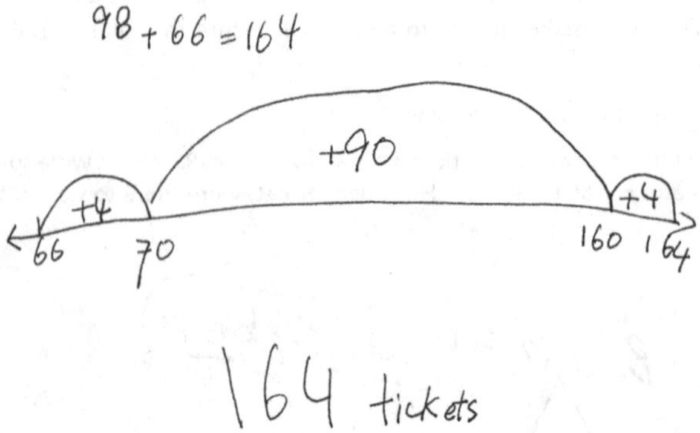

Liam's solution

Base-ten models are not as useful in supporting incrementing strategies. In order to represent the addends with base-ten blocks, the student must begin by decomposing both numbers by place value, and then they are essentially counting all by tens and ones to find the sum (as shown in Figure 6.11). In contrast, the empty number line allows students to keep one addend whole and decompose only the second one to count on by tens and ones.

Starting with one addend and adding the other one in parts is an efficient and generalizable strategy. Unlike decomposition and recomposition, it can be applied to subtraction (as decrementing) without needing to use negative numbers, emphasizing the important relationship between addition and subtraction.

Chapter 7 Developing Whole Number Subtraction for more on decrementing strategies (subtracting one number in parts) and the relationship between addition and subtraction.

Compensation Strategies

Strategies that involve changing one or both addends and adjusting can be very efficient but are dependent on the numbers in the problem. Using these strategies requires that students look at all the quantities to consider whether any of them are close to multiples of ten, hundred, etc. or other familiar landmarks (e.g., 25, 500) before performing the operation.

Compensation strategies involve both an understanding of addition as well as the relationship between addition and subtraction. In Figure 6.17, Elijah changes the problem into 25 + 25 by subtracting 3 from 28. His solution shows an understanding that subtracting 3 from one addend requires adding 3 onto the total.

Figure 6.17 Elijah's subtracts from one addend and adds to the total

Hasan played tag for 25 minutes and then played basketball for 28 minutes. How many minutes did Hasan play altogether?

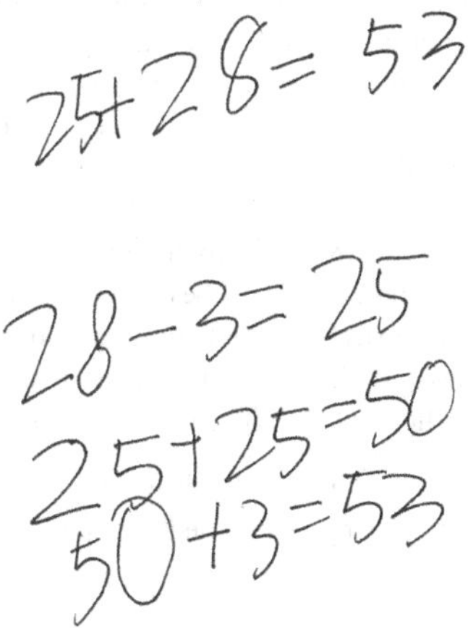

$$25 + 28 = 53$$

$$28 - 3 = 25$$
$$25 + 25 = 50$$
$$50 + 3 = 53$$

The concept of compensation can be supported in the early grades by illustrating combinations of smaller quantities on ten-frames or ten-strips. This is also a useful strategy for deriving single-digit addition facts. Using compensation as a strategy involves judgement on the part of the student as the amount to be added or subtracted depends on the quantities involved and their proximity to landmark numbers (Bass, 2003). The problem in Figure 6.17 provides a good opportunity for the use of compensation because the first addend is a landmark number (25) and the second addend (28) is close to that landmark.

 Chapter 5 Visual Models to Support Additive Reasoning for more on Go To ten-strips and the use of visual models to support flexible strategies and Chapter 9 Developing Math Fact Fluency for more on compensation as a strategy for learning addition facts.

Instructional Strategies to Support the Development of Additive Strategies

In addition to using visual models, there are some effective instructional strategies teachers can use to explicitly introduce and focus on additive strategies while building on student sensemaking and developing understanding. These strategies are important for ensuring equity and access to additive reasoning for all students. We focus here on two strategies for incorporating this instruction into number talk routines: (1) number strings and (2) true/false and open-ended number sentences.

Number Strings

Teachers can support the development of all three of these types of strategies by regular use of mental math or number talk routines that uses a strategically chosen set of problems or *number strings* to scaffold and highlight a strategy (Bray & Maldonado, 2018; Fosnot & Dolk, 2001; Lambert, Imm, & Williams, 2017). The number string routine involves the following steps:

- The teacher writes one problem on the board and then asks students to solve it mentally.
- After collecting different answers from students, without evaluating the correctness, the teacher asks students to justify their answers.
- As students describe their strategies, the teacher asks clarifying and probing questions and then represents the strategy on the board with either equations or visual models so that other students can access the strategy.
- The teacher may also ask students to explain another student's strategy, compare two strategies, or explain why a strategy works.
- Carefully selected questions and probes are used to engage students in making sense of different strategies and come to a consensus on the answer.
- After a few strategies are shared, and there is agreement on the answer, the teacher presents the next problem in the string.

An important feature of number strings is the way they are constructed to scaffold the use of a strategy. The following string is designed to support a compensation strategy:

20 + 14
19 + 14
25 + 9
36 + 19

The first problem is one that children are likely to be able to solve mentally, by using either a decomposing and recomposing strategy or an incrementing strategy. The second problem is related; the first addend is only one less than in the previous problem. Since students will have already agreed that the sum of 20 and 14 is 34, they are likely to notice that the sum of 19 and 14 will be one less than 34. The teacher can ask a student to explain that idea and then represent it in a way that other students can make sense of it. See Figure 6.18 for an example.

Figure 6.18 A teacher's representation of the relationship between two problems in a number string highlights compensation

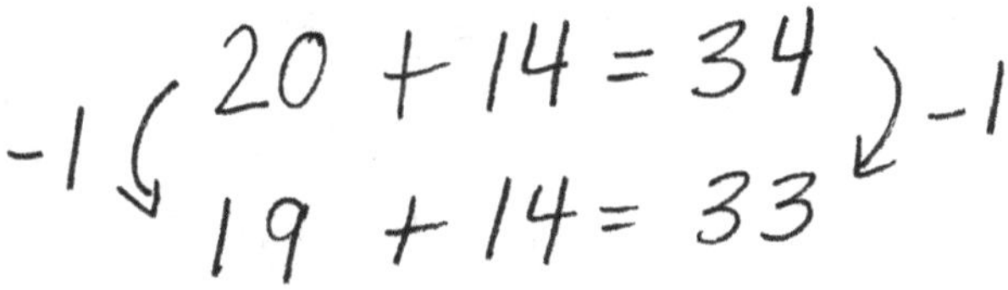

$$-1 \left(\begin{matrix} 20 + 14 = 34 \\ 19 + 14 = 33 \end{matrix} \right) -1$$

The last two problems in the string are constructed to allow students to apply this strategy in increasingly challenging problems.

It is important to note that while the strings are designed to highlight a strategy, students are always encouraged to make sense of the problem and use whatever strategy they choose to find the answer. They are not asked or expected to use a particular strategy; rather, because multiple strategies are elicited and represented, children are exposed to different ideas and have an opportunity to see the connections between different approaches. The discussion helps to scaffold and support the use of increasingly sophisticated or efficient strategies. Table 6.2 shows some sample number strings designed to highlight the three main types of addition strategies. As you look over these strings, note how the strings begin with a problem that would be accessible for most students and how the relationship between the first and second problems set up a strategy that can be used to solve the subsequent problems.

Table 6.2 Sample Number Strings to support *Additive* strategies

Decomposing and Recomposing	Incrementing	Compensation
40 + 50	34 + 10	7 + 7
42 + 53	34 + 13	6 + 8
35 + 22	34 + 23	9 + 6
37 + 68	54 + 35	7 + 9

There are many printed resources and websites to support the use of this instructional activity (For example, see Teacher Education by Design (TEDD), n.d.

True–False Equations and Open Number Sentences

True–False equations or open number sentences can also be used in number talks to support student reasoning about the concept of compensation and properties of addition. Notice the difference in the written solutions to the open number sentence shown in Figure 6.19.

Figure 6.19 An open number sentence focused on compensation

Put a number on the line that makes the equation true. Show or explain how you know.

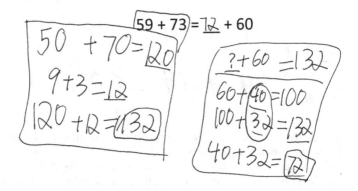

$$59 + 73 = \underline{72} + 60$$

$$50 + 70 = 120$$
$$9 + 3 = 12$$
$$120 + 12 = 132$$

$$? + 60 = 132$$
$$60 + 40 = 100$$
$$100 + 32 = 132$$
$$40 + 32 = 72$$

Lyle's solution

$$59 + 73 = \underline{72} + 60$$

60 is 1 more than 59
So you subtract one from 73.

Hillary's solution

Lyle uses his understanding of equality to solve the problem by first adding up the total of 59 and 73 to get 132 and then adding up from 60 to 132 to find the missing quantity. In the second solution, Hillary notices that since 60 is one more than 59, the number in the blank must be one less than 73. She does not need to add or subtract the quantities to know that this compensation will keep the two sides of the equation balanced. Notably, she is thinking about the relationship between the quantities rather than only performing calculations, an important goal of additive reasoning. Students without an understanding of equality will often add 59 and 73 and write 132 on the line, as if the equal sign means "the answer is." These types of problems can be used to help students develop an understanding of relational thinking and equality (Carpenter, Franke, & Levi, 2003).

 Try asking problems like the following during number talks to focus on understanding of equality, compensation, and properties of addition. Highlight student strategies that use relationships rather than computation.

- 22 + 18 = 18 + 22 (True or false?)
- 68 + □ = 57 + 69
- 13 + 24 + 42 = □ + 42 + 23
- 4 + 5 + 6 = 10 + 5 (True or false?)
- 59 + 36 + 1 = 60 + 36 (True or false?)

Traditional and Transparent Algorithms

The addition strategies described in the above sections develop from, and alongside, students' growing understanding of base-ten number concepts. An *algorithm* is a learned procedure or set of efficient steps for solving a problem. Standard and traditional algorithms are not invented by students but can be related to their derived strategies through intentional instruction.

Look at the three different student solutions to a multi-digit addition problem in Figure 6.20.

- What connections do you see between these three solutions?
- How could you help students see those connections?

Figure 6.20 Three solutions that show evidence of related addition strategies

A movie theater was selling tickets for a new movie. 98 tickets were sold, and now the theater has 66 tickets left. How many tickets were there to start with?

Prita's solution:

Decompose and recompose by place value strategy

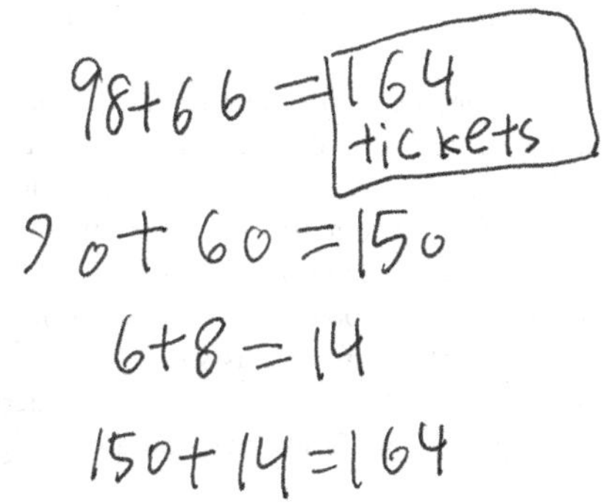

$$98 + 66 = \boxed{164 \text{ tickets}}$$
$$90 + 60 = 150$$
$$6 + 8 = 14$$
$$150 + 14 = 164$$

Figure 6.20 Continued.

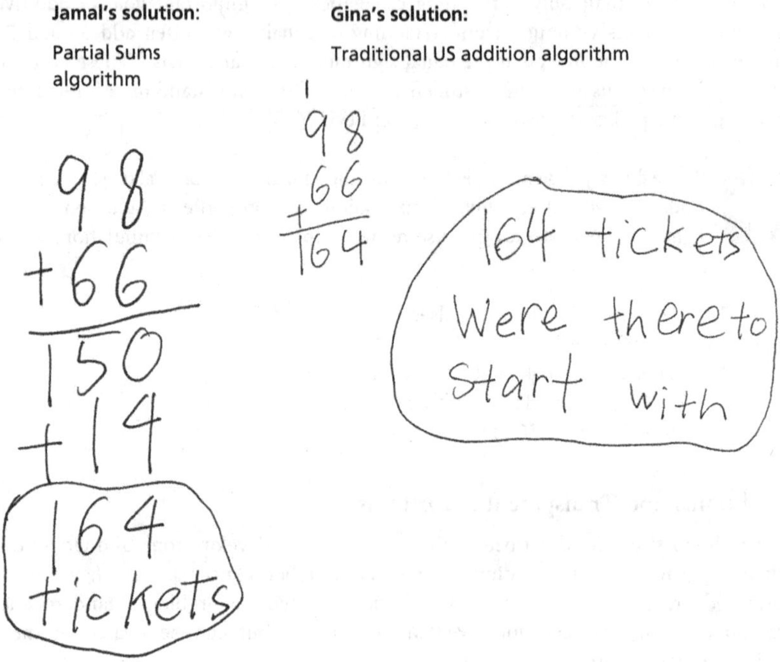

Jamal's solution:

Partial Sums
algorithm

Gina's solution:

Traditional US addition algorithm

$$
\begin{array}{r}
9\ 8 \\
+\ 6\ 6 \\
\hline
1\ 5\ 0 \\
+\ 1\ 4 \\
\hline
1\ 6\ 4
\end{array}
$$

tickets

$$
\begin{array}{r}
{}^{1}\ \ \\
9\ 8 \\
+\ 6\ 6 \\
\hline
1\ 6\ 4
\end{array}
$$

164 tickets were there to start with

Prita's solution illustrates a decomposing and recomposing by place value strategy. She decomposes both quantities into tens and ones, adds the tens and ones separately, and then adds the subtotals. Jamal's solution illustrates the Partial Sums algorithm, a method for efficiently recording the addition of the tens and then the ones, and then finding the total. Note that Prita and Jamal perform the same operations (90 + 60, 6 + 8, and 150 + 14), it is just the written recording that differs.

Gina's solution illustrates the Traditional US algorithm for addition, the procedure most commonly taught in US schools. She begins from the right by adding the digits in the ones place, 6 and 8. Rather than record the sum as 14, she "puts down" the 4 and "carries" the ten to the tens column and records it as a 1 (signifying one ten), adding it to the 9 and 6. Gina is still adding the ones (6 + 8) and the tens (90 + 60), but the 10 from the 6 + 8 is written as a 1 and added to the tens in the second step.

Both algorithms are variations of the decompose and recompose by place value strategy shown in Prita's solution. The Partial Sums algorithm is *transparent*, in that the place value of the quantities is maintained throughout the process (Bass, 2003); children never think of or write a 10 as a 1 (or 90 as a 9). Moreover, each step involves a calculation that can be easily identified. The Traditional US algorithm is also a variation of the decomposing and recomposing strategy; the shortcuts (writing a 10 as a 1 and treating the tens as single digits) make it efficient and "elegantly compact" but at the same time more difficult to understand why it works or makes sense (Bass, 2003, p. 323). As Bass notes, traditional algorithms:

...tend to be cleverly efficient (minimizing the amount of space and writing used) but also opaque (the steps are not notationally expressive of their mathematical meaning). Therefore, if these algorithms are learned mechanically and by rote, the opaque knowledge, unsupported by sensemaking and understanding, often is fragile and error-prone, as many researchers have documented.

(p. 326)

The Traditional US algorithms for addition and subtraction were invented over hundreds of years to enable efficient calculation with pencil and paper. While they are very helpful for handwritten computation with large or multiple quantities, the steps that make them efficient often mask the meaning and place value of the numbers—for example, a ten or a hundred is written and considered as a one during the process. It can often be more difficult to keep track of the steps and therefore less efficient when doing mental calculations. As students add, "put down" or "carry" digits that have been separated from the initial quantities, they often lose sight of the actual value of the quantities and may fail to recognize if they produce an unreasonable answer.

For an example, see Kelsey's solution in Figure 6.21. Kelsey follows all the correct steps of the algorithm, adding single digits and regrouping when necessary, but she does not initially align the quantities by place value; in essence she is adding 132 and 890 instead of 132 and 89. She fails to recognize the importance of place value or to see that the resulting answer is unreasonable. This solution falls under the category of "uses procedures incorrectly" in the *Non-Additive Strategies* section of the *OGAP Addition Progression* and is a common error students make due to lack of transparency of the algorithm.

Figure 6.21 Kelsey uses the Traditional US algorithm for addition without understanding

132 kids went to watch a school play. Then 89 more kids went to watch the school play. How many kids went to watch the school play altogether?

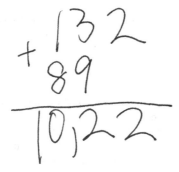

Strategies and Algorithms: Instructional Implications

On the *OGAP Addition Progression*, both strategies and standard algorithms are located on the *Additive* level if they are efficient, accurate, and generalizable. Some of these strategies are not as efficient, however, to use with very large numbers. For example, decomposing a four-digit number by place value results in several parts that must be kept track of and recombined. It therefore becomes important for students to develop an efficient method for recording and keeping track of the steps when they begin to compute with larger numbers in third and fourth grade.

However, there is evidence from research that teaching standard algorithms too early can have detrimental effects on student understanding (Carpenter, Franke, Jacobs, Fennema, & Empson, 1998; Ebby, 2005; Kamii & Dominick, 1998). Kamii and Dominick (1998) argue that algorithms "encourage children to give up their own thinking, and they *unteach* place value, thereby preventing children from developing number sense" (p. 135). Studies by both Kamii and Dominick (1998) and Carpenter et al. (1998) found that students who were not taught algorithms performed better on addition and subtraction problems, getting a larger percentage correct. Moreover, when they did make errors, they tended to be minor calculation errors that still resulted in reasonable answers (i.e., close to the correct answer). On the other hand, students who had been taught algorithms not only made more place value errors but produced incorrect answers that were unreasonable, as evidenced in Kelsey's solution in Figure 6.21. In a case study of one child's problem solving strategies from second to fourth grade, Ebby (2005) found that the student's preference for using algorithms that reduced multidigit problems to a series of one-digit calculations allowed her to continue using her fingers to count by ones as she computed. Although she could often obtain the correct answer to a calculation problem, she did so without having a robust understanding of place value, number sense, or mental strategies, and this impacted her performance in other areas of the mathematics curriculum.

On the other hand, research shows that students who develop their own strategies for addition and subtraction develop understanding of base ten earlier than students who rely on algorithms. In addition, they show more flexibility in applying their understanding to new and more challenging problems and demonstrate fewer errors (Carpenter et al., 1998). As the National Research Council emphasizes, having the freedom to invent strategies that make sense to learners positively affects their overall disposition towards mathematics:

> The invention itself is a kind of problem solving, and they must use reasoning to justify their invented procedure. Students who have invented their own correct procedures also approach mathematics with confidence rather than fear and hesitation
>
> (Kilpatrick et al., 2001, p. 197)

The important question for educators is therefore not whether to teach standard algorithms but when. If standard algorithms are taught before students have developed a strong understanding of base ten and place value, they will think about them as a series of steps to follow. This may also end up delaying their understanding of base ten and number sense. Alternatively, if students are encouraged to first develop and use their own strategies for adding multidigit numbers, their developing understanding of place value and properties of addition can be built upon to transition to the use of standard algorithms. The analysis and comparison of standard computational algorithms, and understanding of why they always work, is an important mathematical goal for the upper elementary grades and beyond (Bass, 2003; Howe & Epp, 2008). This progression from strategies based on understanding to algorithms is reflected in the CCSSM.

Common Core Expectations for Multi-Digit Addition

In the CCSSM, students are expected to develop and use efficient, accurate, and generalizable addition and subtraction *strategies* in grade K–2, before they are expected to use *strategies and algorithms* in grade 3 and then *standard algorithms* in grade 4. As explained in the Common Core Progressions, the standard algorithms can be thought of as the "culmination" of the work done with strategies based on "place value, properties of operations, and/or the relationships between addition and subtraction" in grades K–3 (NGA & CCSSO, 2010).

In grades 1 and 2, students use "concrete models or drawings" as well as "strategies based on place value, properties of operations, and/or the relationship between addition and subtraction" to add within 100 (grade 1) and add and subtract within 1000 (grade 2). There is also a focus on developing the important conceptual foundations for the later use of standard algorithms: "in adding two-digit numbers, one adds tens and tens, ones and ones; and sometimes it is necessary to compose a ten."

In grade 2 the use of strategies is extended to three-digit numbers which adds another level of complexity: composing and decomposing hundreds along with tens and ones. In both grades students are also expected to be able to use reasoning to make connections between strategies, written recordings, and explanations, but in grade 2 there is a specific expectation to use base-ten understanding and properties of operations to "explain why addition and subtraction strategies work."

In grade 3, the major focus shifts to multiplication while continuing to strengthen fluency with addition and subtraction strategies and algorithms based on place value, properties of operations, and/or the relationship between addition and subtraction. At this stage, students should also be able to use efficient written recording methods without the need for concrete models or drawings.

In grade 4, students should be able to fluently add and subtract within 1,000,000 using the standard algorithm. Note that a standard algorithm (Partial Sums or Traditional US) is often more efficient and accurate for numbers of this magnitude as it reduces the number of steps that need to be written out.

In grades 5 and 6, this fluency extends to adding and subtracting decimal numbers, first with concrete models or drawings and strategies based on place value, properties of operations and/or the relationship between addition and subtraction (grade 5) and then with the standard algorithm (grade 6).

In sum, the CCSSM standards for multi-digit addition and subtraction reflect a progression from strategies based on understanding, to fluency with strategies and algorithms, to the use of standard algorithms—as the magnitude of number increases, and base-ten understanding is generalized. Teachers can use the *OGAP Addition Progression*, along with knowledge of the CCSSM, to make decisions about how to use curricular materials. The following vignette describes how one second-grade teacher adapted the lessons in his textbook to better reflect what he knew about how students learn to add multi-digit numbers.

Opening up the Curriculum to Develop Additive Strategies with Understanding

Before teaching the first second-grade unit on multi-digit addition, Mr. Harper looked through his curriculum materials. He noticed that the textbook introduced multiple strategies, focusing on a different strategy in each lesson—including adding ones and tens with 100 charts and base-ten blocks, incrementing by tens on an empty number line, Partial Sums, compensation, and the Traditional US algorithm.

Mr. Harper knew that many of his students were still developing their base-ten understanding and that the standards did not expect students to be able to use a standard algorithm in second grade. He decided to make an important change when he taught this unit. Rather than introducing a new strategy each day and having all students practice that strategy, he would pose the problems that were in the text and allow students to use a strategy that made sense to them. He would make sure to set aside time to have students share and discuss their strategies, so that he could highlight important concepts and help them make connections between strategies. He would also incorporate number talks three times a week to introduce and help students make connections between strategies.

For example, one lesson began with the following problem:

Wes picked 18 pears. Taryn picked 25 pears. How many pears did they pick in all? Solve the problem using Partial Sums. Draw place value blocks to help explain your work.

Mr. Harper decided to use this problem with his students without the provision that it be solved with Partial Sums or place value blocks. After launching the problem to make sure students understood the context, he had the students work independently to solve the problem in their notebooks. As students were working, he observed the strategies they were using, asked probing questions to make sure they were understanding place value of the quantities, and suggested tools such as base-ten blocks or number lines when he thought it would help students make sense of the quantities. He drew on his knowledge of the *OGAP Addition Progression* to take notes on different levels of strategies students were using and make decisions about which strategies, and in what order, to have students share in the whole group discussion.

Based on his observations, Mr. Harper selected three pieces of student work to project on the document camera. He began with Colton's solution shown in Figure 6.22 because he knew several students in the class were using base-ten blocks to decompose and recompose both numbers. He then showed Mabel's solution and asked questions to focus students on the comparison between Colton's representation of 18 and 25 with base-ten blocks and Mabel's use of number bonds.

Figure 6.22 Mrs. Harper started by showing both Colton's and Mabel's solutions

Wes picked 18 pears. Taryn picked 25 pears. How many pears did they pick in all? Show and explain your work.

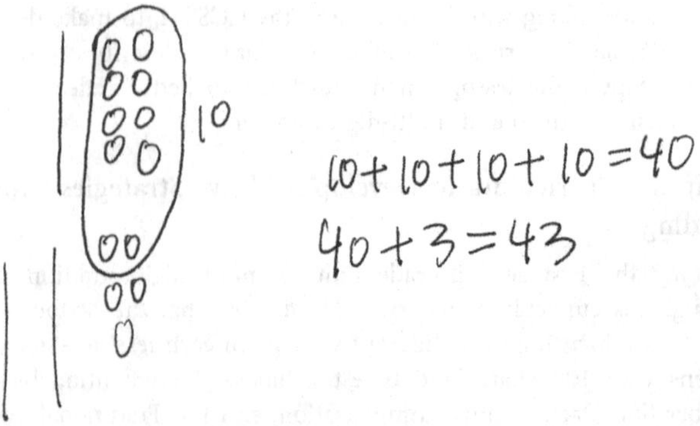

$$10 + 10 + 10 + 10 = 40$$
$$40 + 3 = 43$$

Colton's solution

Figure 6.22 Continued.

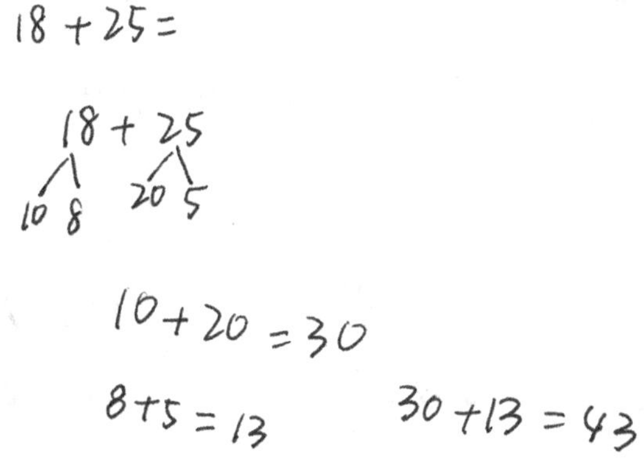

$$18 + 25 =$$

$$18 + 25$$

$$10 \quad 8 \qquad 20 \quad 5$$

$$10 + 20 = 30$$

$$8 + 5 = 13 \qquad 30 + 13 = 43$$

Mabel's solution

Finally, he projected Leyla's solution shown in Figure 6.23 and asked students what was different about the way Leyla solved the problem. The empty number line model helped students see how Leyla started with 18 and then added the 25 on in parts, using jumps of 10. Mr. Harper asked the class if they could think of an easier way to add 20 to 18 and a student suggested taking one jump of 20.

Figure 6.23 Mr. Harper projected Leyla's incrementing strategy using an empty number line

Wes picked 18 pears. Taryn picked 25 pears. How many pears did they pick in all? Show and explain your work.

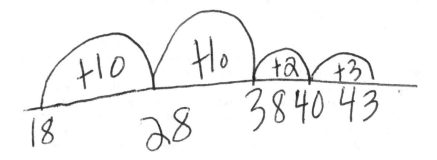

Mr. Harper then used this discussion to launch a mini-lesson on efficient recording of solutions. He demonstrated how Mabel's solution could be recorded vertically using the Partial sums algorithm, writing the decomposition of 18 and 25 next to the algorithm to help students make sense of it. He also showed how

Leyla's solution on the number line could be efficiently recorded, as shown in Figure 6.24. See Ebby, Hulbert, and Fletcher (2019) and Stein and Smith (2018) for more on this strategy of selecting and sequencing student work.

Figure 6.24 Mr. Harper demonstrates efficient recording methods for Mabel's and Leyla's solutions

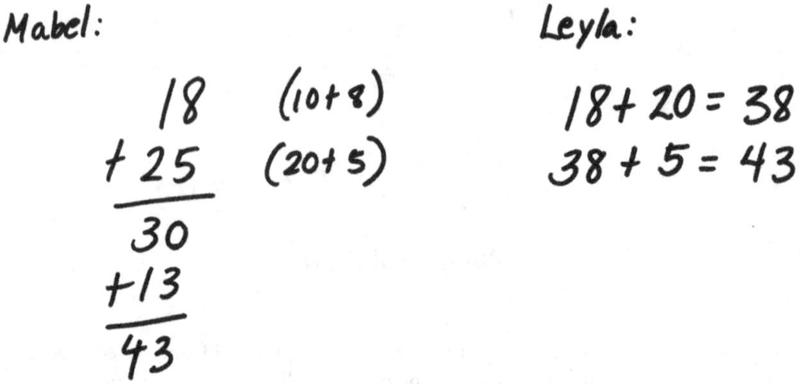

Mabel:

$$18 \quad (10+8)$$
$$+25 \quad (20+5)$$
$$\overline{30}$$
$$+13$$
$$\overline{43}$$

Leyla:

$$18 + 20 = 38$$
$$38 + 5 = 43$$

Mr. Harper told students that these were ways that mathematicians recorded addition, and they could try it out if they wanted. He knew that some students would need to continue to use models, but others could draw on mental models to work with equations and algorithms.

Mr. Harper then purposefully chose a set of practice problems from the text that included a variety of number sizes and additive situations. He encouraged students to choose problems to work on that were "just right" for them in terms of challenge and to try out some new strategies, and then further supported their choices as he circulated. He also paired some students up strategically to work together based on the strategies he had observed them using for the initial problem.

Mr. Harper was pleased with how he was able to allow students to continue using strategies that made sense to them while introducing them to new strategies and a standard algorithm that connected to their developing understanding. He continued to use this strategy to teach the next few lessons in the text, focusing on supporting students to explain their solutions, and building on their strategies to make connections to more efficient strategies. He also regularly gave tasks as exit slips and analyzed student responses with the *OGAP Addition Progression* so that he could track students' use of strategies and underlying issues and errors that needed to be addressed in instruction.

As he progressed through the unit, Mr. Harper decided not to introduce the Traditional US algorithm, even though it was in the text. He was confident that his students had strategies that made sense to them to compute accurately with two-digit numbers, and that they would be able to use those strategies to solve any problems that would be on the state standardized test.

Chapter Summary

The *Additive* level of the *OGAP Addition Progression* includes algorithms and strategies that are efficient and generalizable and are built upon a solid understanding of number, properties, and relationships. There are three main types of strategies for multi-digit addition:

- decomposing and recomposing by place value,
- incrementing, and
- changing one or both numbers and compensating.

Students should have opportunities to develop and use all three types of strategies so that they can deepen their understanding of base ten and number relationships and approach problems with flexibility. The Traditional US and Partial Sums algorithms are variations in ways to perform and record a decomposing and recomposing strategy, and it is important for students to see these connections when they are introduced in instruction.

In the CCSSM, students are not expected to be able to use standard algorithms until third and fourth grade. Students should have experiences developing and using strategies based on understanding of place value and properties of addition before they are introduced to these algorithms.

Looking Back

1. **Building Understanding of Additive Strategies:** Review the *OGAP Additive Framework*. How does the information in the *OGAP Base Ten Number Progression* (p. 4) relate to the strategies students use for addition illustrated on the *OGAP Addition Progression* (p. 2)?

2. **Practice an *OGAP Sort*:** Use the *OGAP Addition Progression* to analyze student thinking and strategies evident in the following samples of student work shown in Figure 6.25. You may want to record the information on an *OGAP Evidence Collection Sheet* (Chapter 2, Figure 2.22).

   (a) For each piece of student work record:

   - the level on the progression the work best reflects
   - any underlying issues or errors
   - the correctness of the solution.

   (b) Based on the analysis of student evidence, how might instruction be designed in the next days to support these four students during whole group instruction?

Figure 6.25 Four student responses to an additive problem

Hasan played tag for 25 minutes and then played basketball for 28 minutes. How many minutes did Hasan play altogether?

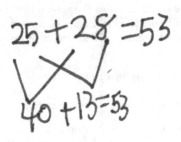

$$25 + 28 = 53$$

$$40 + 13 = 53$$

Student A

$$10 \quad 20 \quad 30 \quad 40$$

$$\begin{array}{cccccccccc} & & & & 1 & 2 & 3 & 4 & 5 \\ & & & & & & & & & 6 & 7 & 8 & 9 & 10 & 11 & 12 & 13 \end{array}$$

$$10 + 10 + 10 + 10 + 8 + 5 = 53$$

$$25 + 28 = \underline{53} \text{ minutes}$$

Student B

$$25 + 28 = 58$$

Student C

Figure 6.25 Continued.

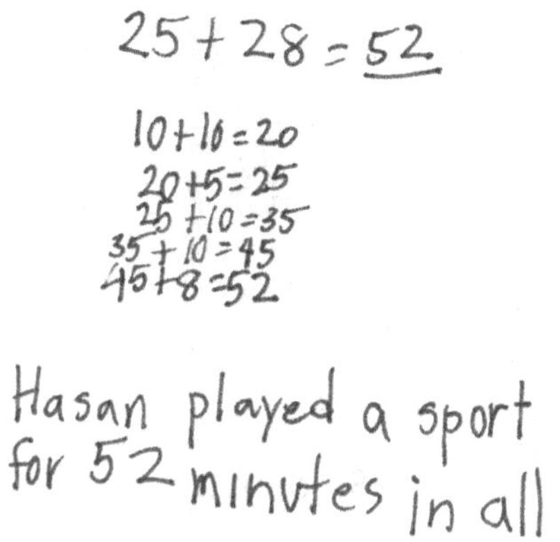
25 + 28 = 52
10+10=20
20+5=25
25+10=35
35+10=45
45+8=52
Hasan played a sport for 52 minutes in all

Student D

3. **Base-Ten Block Models and Representations:** The use of base-ten block models results in a strategy that is essentially a counting all in units of tens and ones. This chapter presented three student responses with base-ten representations that are presented here again in Figure 6.26.

Figure 6.26 Three student responses using base-ten block representations

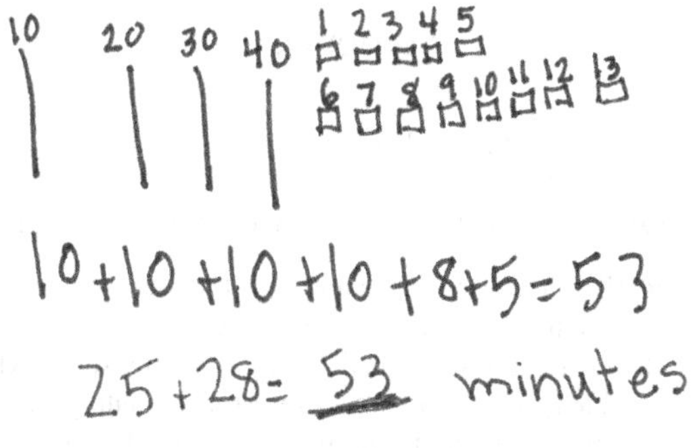
10 20 30 40
1 2 3 4 5
6 7 8 9 10 11 12 13
10+10+10+10+8+5=53
25+28= 53 minutes

Student A

Figure 6.26 Continued.

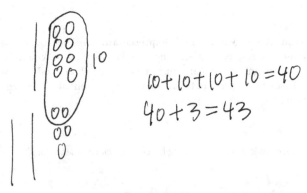

$$10+10+10+10+10=50$$
$$50+3=53$$

Student B

$$10+10+10+10=40$$
$$40+3=43$$

Student C

(a) What evidence in the student responses illustrates a counting-all-by-units strategy?

(b) What features of base-ten models make them a limiting strategy?

4. **A Classroom Vignette:** Mr. Capps and his students are working in a unit on addition and subtraction. He notices most of his students make use of base-ten number concepts, properties, and relationships when solving a variety of problem types. Further, many students efficiently record their thinking symbolically without the use of a visual model. Figure 6.27 shows the responses of three of Mr. Capp's students.

Figure 6.27 Three student solutions from Mr. Capp's class

Reese collected 352 pennies.

She collected 129 fewer pennies than Paula.

How many pennies did Paula collect?

Show or explain how you know.

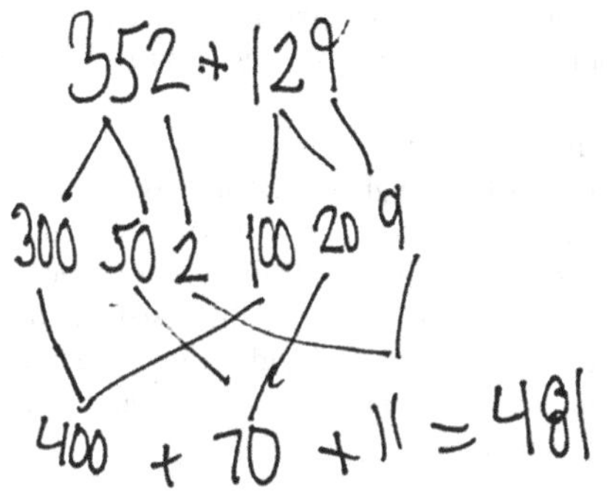

$$352 + 129$$

300 50 2 100 20 9

$$400 + 70 + 11 = 481$$

Student A

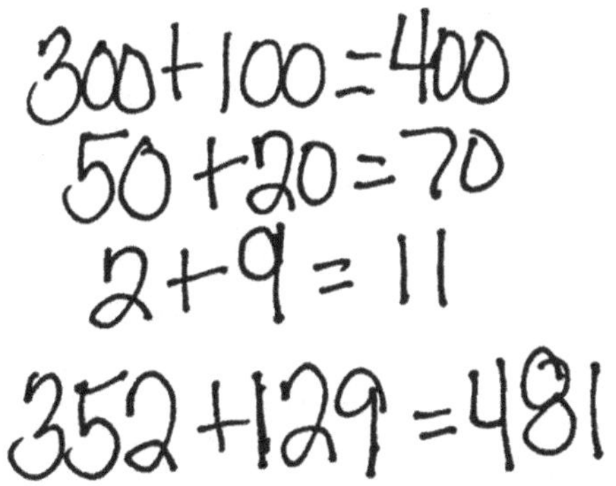

$$300 + 100 = 400$$
$$50 + 20 = 70$$
$$2 + 9 = 11$$

$$352 + 129 = 481$$

Student B

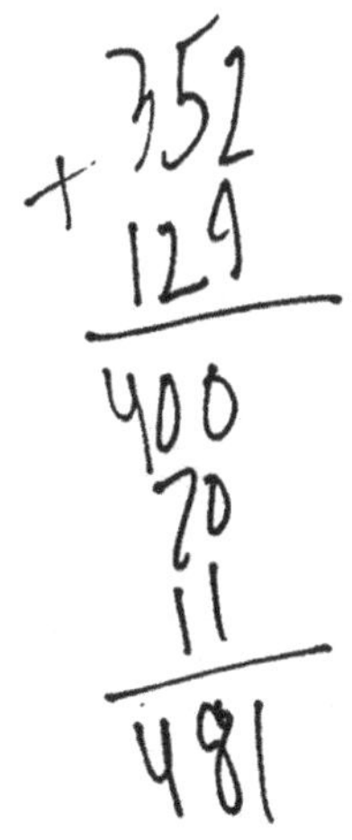

$$\begin{array}{r} 352 \\ + 129 \\ \hline 400 \\ 70 \\ 11 \\ \hline 481 \end{array}$$

Student C

(a) What level of the *OGAP Addition Progression* best describes each response?

(b) What additive strategy type (Table 6.1) describes the three responses?

(c) How does the recording used by Student B serve as a bridge between the recording of Student A and Student C?

Instructional Link

Use the following questions and Table 6.2 to consider how your instruction and math program offer opportunities for students to share strategies and make connections to strategies shared by others.

1. To what extent are students presented with opportunities to develop and share their own *strategies* for addition in whole group, small group, or partner settings?
2. To what degree do your math curriculum materials align with developmentally appropriate exposure to *strategies* and *algorithms* for the grade you teach?
3. Consider gathering Number Talk and Number String resources to augment instruction. Work through each string you consider and identify the scaffolded strategy, possible student responses, and how anticipated responses could be recorded (scribed) during discussion.
4. Try It: Select a strategy from the types discussed (decompose and recompose, incrementing, compensation) and create a number string appropriate for the grade you teach. Share it with a fellow teacher and anticipate possible student responses.

7
Developing Whole Number Subtraction

Big Ideas

- Subtraction is a more complex operation than addition.
- Understanding and using the relationship between addition and subtraction can make subtraction more accessible to students.
- The use of visual models to solve problems can help students make sense of the operation of subtraction.
- Subtraction strategies and algorithms should develop from instruction that begins with conceptual understanding.
- There are several effective strategies and algorithms for solving subtraction problems.

This chapter focuses on the development of strategies and algorithms for subtraction. Subtraction is often more complex than addition, and as a result, students may struggle to use effective strategies to calculate answers accurately (Fuson, 1984). Incorporating a variety of sensemaking strategies, visual models, and a focus on the relationship between addition and subtraction can make subtraction significantly more accessible to children.

While young children's first experiences with subtraction are often about taking away, for students to have a complete understanding of subtraction they must expand understanding to include a difference (or distance) meaning. Chapter 8 will discuss how this understanding relates to problem structures, but in this chapter we will examine these meanings in relation to strategies and algorithms.

The *OGAP Subtraction Progression* illustrates the development of subtraction strategies, beginning with counting all, counting back, and counting on at the earlier stages of development. Once students understand and can use the relationship between addition and subtraction along with developing base-ten understanding, they can make use of visual models to move towards fluency with subtraction of larger numbers. At the top of the *OGAP Subtraction Progression* are *Additive Strategies* which include strategies and algorithms that are not taught procedurally or in isolation but rather built upon understanding that has been developed through visual models and base-ten understanding. When procedures are connected to the related underlying concepts, students have a better chance of using procedures appropriately and extending them to unfamiliar situations (Fuson, Kalchman, & Bransford, 2005). As with addition, strategies and algorithms for subtraction should be efficient, accurate, and generalizable towards the goal of computational fluency (Russell, 2000).

Early Strategies for Subtraction

Early learners often first understand subtraction as a take-away situation and they solve problems by removing objects from the total and counting what remains. Consider the following problem:

> 7 ladybugs were on a leaf and then 3 ladybugs crawled away. How many ladybugs are left?

An early strategy for solving this task would be to count out 7 items, count 3 out of the 7, pull them away, and then count what is left, essentially treating subtraction as a counting exercise. This strategy allows students to count forward to take away and avoids the need for students to count back, which is a significantly more difficult process. However, it is a cumbersome and inefficient strategy that involves counting three times.

In Figure 7.1 Sara uses the direct modeling strategy of count all to solve a problem with two-digit quantities. She draws all 32 crayons with circles and takes away by counting forward and crossing out 17 of them. She then circles what she has left in groups of 5 to generate the correct solution of 15. On the *OGAP Subtraction Progression* this work falls in the *Early Counting* level because she is modeling all quantities and then counting what she took away as well as what remains.

Figure 7.1 Sara's *Early Counting* strategy

> Juan has 32 crayons. 17 crayons are in his desk. The rest are in his backpack. How many crayons are in his backpack?

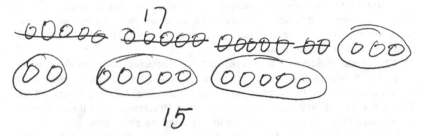

In Figure 7.2 Ana and Simeon also used counting strategies to solve the same problem but neither of them needed to model all or count all. Ana was able to think of the task as related to addition, counting on by ones from 17 to 32 to arrive at her answer of 15. She keeps track of the amount she is counting on with a separate counting chain above the numbers and indicates that she is thinking of this as addition by writing an equation with a missing addend.

Simeon does not need to model 32 and instead counts back from 32. He also appears to estimate that he does not need to go past 13 when writing his numbers. The dots above his numbers indicate that he counts back 17 to arrive at 15. Although the students solved the task differently, both solutions would fall in the *Counting* level of the *OGAP Subtraction Progression*.

Figure 7.2 Ana and Simeon's counting strategies

Juan has 32 crayons. 17 crayons are in his desk. The rest are in his backpack. How many crayons are in his backpack?

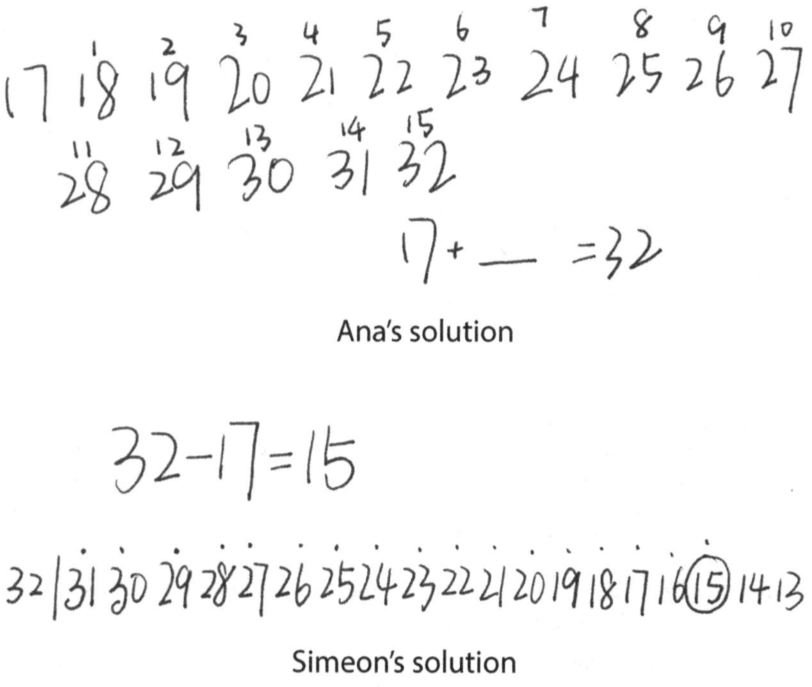

Ana's solution

Simeon's solution

For young children counting back is more difficult than counting forward. While counting forward is easier, it can be confusing for children to understand how counting up, like Sara and Ana did in the work in Figures 7.1 and 7.2, can be considered taking away (Baroody, Ginsburg, & Waxman, 1983; Fuson, 1986). Reading an equation such as 7 – 3 = 4 as "7 minus 3 equals 4" instead of "7 take away 3 equals 4" is one way to help students avoid generalizing subtraction as take away and supports the extension of adding up to subtract, thus supporting both Ana and Simeon's strategies.

While students may rely on counting as a method that can be successful with smaller numbers, as numbers become larger and problem types are varied, these counting strategies will be time consuming, more difficult, and less useful as an effective strategy. In Figure 7.3 Finn uses an *Early Counting* strategy with a task that requires solving 56 – 27. He writes all the numbers from 56 to 27 and then counts the numbers in between resulting in an incorrect solution. His work indicates that he may be counting the numbers between the quantities rather than thinking about the distance between 56 and 27, which could lead to some challenges as he transitions to the use of a number line. Students often struggle to make sense of what is actually happening when using counting strategies to solve problems with larger numbers because they have to focus on keeping track of so many quantities.

Figure 7.3 Finn's counting on solution shows a common misconception about subtraction

A string was 56 inches long. Dylan cut some off. Now the string is 27 inches long. How much of the string did Dylan cut off?

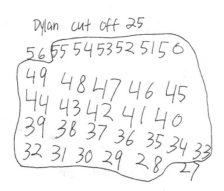

With increased understanding of properties of addition, number relationships, and base-ten fluency, students can use more flexible and efficient strategies, decreasing the likelihood that errors will occur. It is hard to imagine a student solving a problem such as 293 – 136 by using counting strategies accurately given the amount of work, time, and organization involved to carry out the strategy successfully.

Understanding Properties of Addition and the Relationship between Addition and Subtraction

An understanding of the commutative and associative properties is very helpful for students in developing flexibility with addition. Since addition is commutative, when adding 3 + 8, whether a student starts with 8 and adds 3 on or starts with 3 and adds 8 on makes no difference. However, in subtraction this move will result in an incorrect solution. As Fuson et al. (1997) state, this lack of commutativity is "one inherent source of difficulty in subtraction" (p. 151).

Thus, part of understanding the properties of addition is understanding that they do not extend similarly to subtraction. This idea is reliant on students having a strong understanding of the relationship of numbers in addition and subtraction equations. The most important relationships for supporting flexibility in subtraction include the following ideas:

- Since subtraction is the inverse of addition, one can use addition to solve subtraction. Helping students see subtraction as a "think addition" strategy opens up many opportunities by using an easier operation to solve a much more difficult one.

- The numbers in a subtraction equation do not represent the same thing as in addition and cannot be manipulated and adjusted in the same ways. In addition equations, the addends can be considered parts and the answer is the whole; in subtraction equations the first number is the whole, the other number is the part, and the answer is the other part. See Figure 7.4 for an illustration of this idea.

Figure 7.4 A part–part–whole model has been used to illustrate the difference in the meaning of the numbers in the operations of addition and subtraction

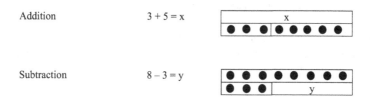

Math education researchers have proposed that many of the challenges associated with subtraction can be offset by teaching addition and subtraction concurrently; otherwise learners are more likely to generalize attributes of addition to subtraction. This is particularly true when children's instruction is focused on addition for a long time before they learn subtraction (Fuson et al., 1997). Students' ability to understand and use the relationship between addition and subtraction is an indication of their developing additive reasoning, as well as evidence of understanding that can be built upon to move students to more sophisticated strategies.

Using true–false questions as warm-ups or number talks with the class is one way that teachers can help to expose or prevent misconceptions about these relationships and properties. Teachers might present a task such as those below, ask students to solve and justify their answers, monitor student thinking for understandings and misunderstandings, and discuss them as an entire class or in small groups.

 Questions to pose to discuss subtraction properties and relationships.

- $7 - 1 = 1 - 7$ (True or false?)
- $8 + 6 = 6 + 8$ so $8 - 6 = 6 - 8$ (True or false)?
- $13 - 8 = \_\_$. Billy wrote $8 + \_\_ = 13$ to find the answer. Can this work? Explain.

 Chapter 6 Developing Whole Number Addition for more on properties of addition and the routine of number talks.

Visual Models to Support Understanding of Subtraction

Visual models play an integral role in developing fluency with subtraction, as shown on the *Transitional* level of the *OGAP Subtraction Progression*. Students who can use flexible visual models for making sense of subtraction often have an advantage over students who do not posess the same knowledge. Additionally, once students can use their understanding of unitizing and base-ten number composition, they can subtract using place value units rather than only ones. This will result in more efficient strategies that have less room for error.

In Figure 7.5 Micah and Rico both have solutions that fall in the *Early Transitional* level of the *OGAP Subtraction Progression*. Examine the two strategies in Figure 7.5. How are they similar and how do they differ?

Figure 7.5 Micah and Rico's *Early Transitional* strategies

Javier drove 38 miles. Shrina drove 101 miles. How many fewer miles did Javier drive than Shrina?

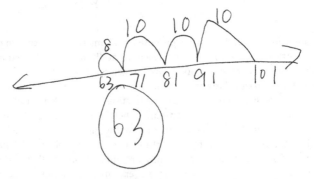

Micah's solution

Rico's solution

The solutions in Figure 7.5 share many similarities and a few distinct differences. Both Micah and Rico use their developing understanding of base ten to subtract 38 from 101 by subtracting 3 tens and 8 ones to arrive at a correct answer. They both solve the task using a take-away strategy versus an add-up strategy, and they both use a strategy that requires they keep track of the 38 they are taking away.

Micah sketches a number line, starts at 101, jumps back by 10s off the decade until he arrives at 71, and then jumps back by 8 to get an answer of 63. The numbers in his problem are easily seen in his solution: the 101 is where he starts, the 38 is the distance from 101 to 63 (likely thought of as taking away 38 from 101), and the 63 is where he arrives on the number line after taking the 38 away.

Rico sketches base-ten blocks to solve the task. He draws ten ten-blocks and one one-block to represent the 101 miles Shrina drove. The original model Rico drew is less easy to see since he must cross some out and trade a ten-block for ten ones, as he does to the right of his sketch. This is one reason base-ten blocks can be a difficult model for students to use: the model does not maintain itself and must be altered to make the subtraction possible. For students with working memory limitations, this change in the model can add one more challenge to an already challenging task. Rico now crosses out 38 (3 ten-blocks and 8 ones) and then counts up what is left.

When considering the use of base-ten blocks for multi-digit addition and subtraction, teachers must be mindful of the limitations they present. The inflexible nature of the model results in students needing to "trade" in order to get the blocks in a form that can be manipulated to derive the answer. To use the base-ten blocks Rico had to count out 101 blocks, count out ten ones to trade for a ten, count out the 3 tens and then the 8 ones, and then count what was left. While base-ten blocks offer a visual of place value units, an improvement over counting by ones, there is still a significant amount of counting that occurs to find the answer. This results in a strategy focused on a count all, count back, count what is left approach to subtraction with much room for error and a challenge for keeping track throughout the process.

For this reason, base-ten models, which represent every ten, appear in the *Early Transitional* section of the *OGAP Subtraction Progression*. Empty number lines, on the other hand, can support students to work with multiples of tens, as illustrated in Dani's solution in Figure 7.6. Dani solves the same task as Rico and Micah did in Figure 7.5, but the evidence in the work shows that Dani can flexibly decompose the 38 into 1 + 30 + 7. As a result, she arrives at the correct solution of 63 more efficiently than the other two students. The ability to decompose in ways that support the numbers and make the task more efficient is evidence of a student with strong number sense. In Dani's case, she took 1 from 101 to reach the 100 making it much easier to take 30 away in one chunk. Empty number lines support the use of these more flexible and efficient approaches to subtraction.

Figure 7.6 Dani's *Transitional* strategy

Javier drove 38 miles. Shrina drove 101 miles. How many fewer miles did Javier drive than Shrina?

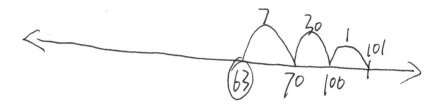

See Chapter 5 Visual Models to Support Additive Reasoning for more on number lines.

When students' work shows evidence of *Transitional Strategies*, their work shows they are using their base-ten understanding and efficient models to solve a variety of subtraction problems. One benefit of these strategies is that while learning how to subtract, students are also focusing on fluency with base ten, resulting in more sense-making and more reasonable solutions. In Figures 7.7 and 7.8 below Drew and Gia use *Transitional Strategies* to solve the same task. Their base-ten understanding is visible in both solutions.

Drew solves the task as a missing addend problem, adding up from 53 to 97 by finding how far away the two numbers are from each other, or their difference. He is able to jump 40 off of 53 to 93 and then add four more to get to 97, arriving at a solution of 44. This adding up strategy allows a student to place both 53 and 97 on the number line and find the difference without needing to keep track along the way. In the end the jumps can simply be added together.

Figure 7.7 Drew's *Transitional* strategy. Drew finds the difference between 53 and 97

Juan drove from Philadelphia to New York. He left Philadelphia, drove 53 miles, and then stopped to eat lunch. After lunch, he drove the rest of the way to New York. He drove 97 miles altogether. How far did Juan drive after lunch?

Gia takes a slightly different approach with her number line model. She begins her empty number line at zero and indicates that she takes 53 away from the beginning of the number line by crossing out the distance from 0 to 53, rather than counting back from 97. This also results in her adding up to see what she has left between 53 and 97, taking one big jump of 40 and then 4 more. Since the distance from 0 to 97 includes the 53, it can be taken away from 97 anywhere along that distance. The idea of taking away from the beginning of the number line versus counting back from the end of the number has been recommended by some researchers since counting up is easier than counting back (Fuson, 1986). Additionally, exposing students to both methods illustrated in Figures 7.7 and 7.8 can help them make connections between different meanings for subtraction, as discussed in the next section.

Figure 7.8 Gia's *Transitional* strategy. Gia takes away 53 from the beginning of the number line

Juan drove from Philadelphia to New York. He left Philadelphia, drove 53 miles, and then stopped to eat lunch. After lunch, he drove the rest of the way to New York. He drove 97 miles altogether. How far did Juan drive after lunch?

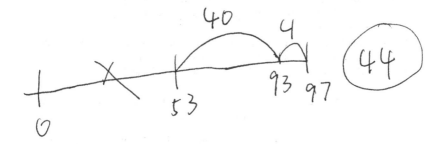

Subtraction as Take Away and Difference

There are two distinct meanings of subtraction: take-away and difference. Young children are generally first exposed to subtraction as a take-away: starting with one quantity and taking some away to arrive at the answer. A different way to conceptualize subtraction is to find the difference in value between two quantities. Effective use of visual models, strategies, and algorithms can help expand students' understanding of subtraction to include difference.

Empty number lines can support both meanings of subtraction. In Figure 7.6 Dani takes 38 away from 101 by counting back on the number line. Her answer is where she lands on the number line, and her strategy indicates she thought about the task as taking away. In Figure 7.7 Drew placed 97 and 53 on the number line and found how far they were from each other. The answer to the problem can be found in the distance, or difference between the two numbers. Both students used the empty number line but thought about subtraction differently. In addition to models, number-based strategies, and algorithms also rely primarily on one meaning for subtraction or the other.

Additive Strategies and Algorithms for Subtraction

As mentioned in Chapter 6, students can invent or develop strategies for multi-digit operations without being directly taught procedures. Researchers have found that students almost always begin with the largest units when using these invented methods, moving left to right to compute, and using base-ten understanding as well as the relationship between addition and subtraction (Fuson et al., 1997; Kamii, 1989). Research by Carpenter and colleagues (1998) also shows that "children who use invented strategies develop knowledge of base-ten concepts earlier than children who rely more on standard algorithms" (p. 46).

As discussed in more detail in Chapter 6, algorithms are a set of efficient steps that are carried out for solving a problem and most often are taught, not student-generated. Both strategies and algorithms are represented at the top of the *OGAP Subtraction Progression* as long as they are efficient, accurate, and generalizable. It is important to remember that when students have learning experiences focused on developing strategies with understanding it can lead to greater sensemaking when exposed to algorithms.

Below we will discuss both strategies and algorithms, but we begin with some information about strategies for subtraction. As with addition, while there are many variations, subtraction strategies fall within some general categories (Fuson et al., 1997):

1. **Decompose and Recompose by Place Value.** Break up the numbers by place value parts, subtract the place value parts or units separately, and then combine the subtotals.
2. **Decrement by place value.** Keep the minuend (the part you start with) whole and take away the subtrahend using place value parts.
3. **Find the Difference.** Keep both numbers whole and add or subtract mainly by place value parts to find the difference.
4. **Constant Difference.** Change one or both numbers to make an easier problem and then adjust if needed to maintain the difference.

Table 7.1 provides an overview of these subtraction strategies with examples of each when used with two and three-digit numbers. These strategies and related algorithms will each then be discussed in more detail.

Table 7.1 Additive strategies for two and three-digit subtraction

Strategy	Description	Two-digit example: 53 – 27	Three-digit example 578 – 139
Decompose and recompose by place value	Break up both numbers by place value, subtract by place value	50 – 20 = 30 3 – 7 = -4 30 – 4 = 26	500 – 100 = 400 70 – 30 = 40 8 – 9 = -1 400 + 40 – 1 = 439
Decrement by place value	Keep minuend whole and subtract the subtrahend in primarily place value parts	53 – 20 = 33 33 – 7 = 26	578 – 100 = 478 478 – 30 = 448 448 – 9 = 439
Find the difference	Keep both numbers whole and add or subtract to find the difference mainly by place value parts	27 + 3 = 30 30 + 20 = 50 50 + 3 = 53 3 + 20 + 3 = 26 53 – 20 = 33 33 – 6 = 27 20 + 6 = 26	139 + 400 = 539 539 + 30 = 569 569 + 9 = 578 400 + 30 + 9 = 439
Constant difference	Adjust both numbers to make an easier problem and then find the difference.	56 – 30 = 26 ↓ 53 – 27 = 26	579 – 140 = 439 ↓ 578 – 139 = 439

Decompose and Recompose Strategies

In Figure 7.9 Mason solves a two-digit subtraction problem by decomposing both numbers by place value parts, subtracting by place value, and then combining the differences to arrive at the answer of 29. His understanding of base ten, as well as his ability to use it to subtract, is evident in the work. He subtracts 7 from 6, getting −1 and then takes 1 from the difference of 50 and 20. This strategy is located in the *Additive* level of the *OGAP Subtraction Progression*.

Figure 7.9 Mason decomposes and recomposes by place value

A string was 56 inches long. Dylan cut some off. Now the string is 27 inches long. How much of the string did Dylan cut off?

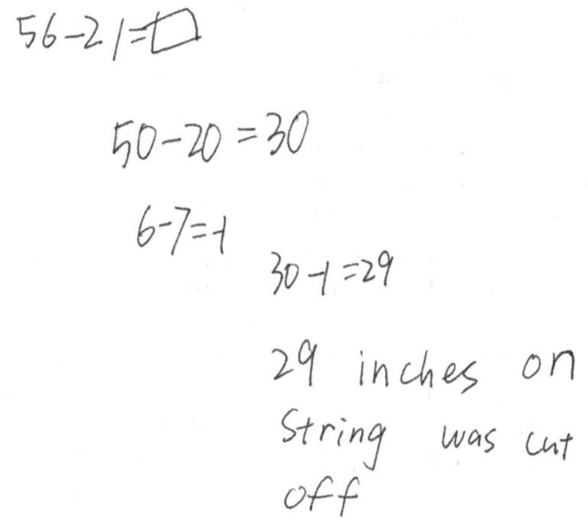

$$56-21=\square$$

$$50-20=30$$

$$6-7=-1$$

$$30-1=29$$

29 inches on String was cut off

One of the benefits of this strategy is that it is can be efficient when used as an algorithm with larger numbers as seen in Figure 7.10. DeMarcus successfully uses the Partial Differences algorithm to find the difference between two three-digit numbers. As in Mason's work, DeMarcus is able to decompose and subtract by place value parts with an understanding of what happens when you do not have enough to take the subtrahend from the minuend in any base-ten position.

Figure 7.10 DeMarcus uses Partial Differences to find the difference between 765 and 398

Kids baked 765 cupcakes for the school bake sale. They sold some cupcakes and 398 are left. How many cupcakes did they sell?

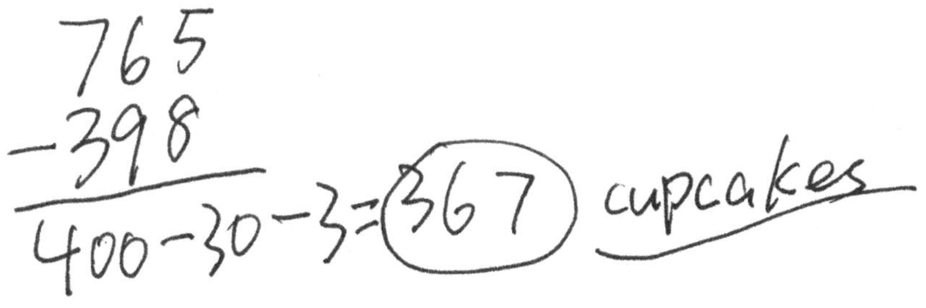

$$765$$
$$-398$$
$$400-30-3=367 \text{ cupcakes}$$

The Partial Differences algorithm can be efficient and easy to use both on paper and as a mental math strategy; however, it requires students understand the idea of negative amounts when taking more away than you start with. For example, when Mason found the difference of 6 and 7 as –1 and not 1, it is evidence that he understood and made sense of the number relationship in the operation. Most students will not naturally come to this notation without instruction, but they may come to the conceptual understanding on their own. Partial Differences is an efficient take-away strategy that offers an advantage in that all the same kinds of steps are taken at the same time: subtract, subtract, combine.

Decrementing by Place Value

Micah's strategy in Figure 7.5 and Dani's strategy in Figure 7.6 both use number lines to take the subtrahend from the minuend, keeping the total whole and decrementing the subtrahend and subtracting primarily by place value parts. Both these strategies fall within the *Transitional* level of the *OGAP Subtraction Progression* as both students use a model to carry out the task.

In Figure 7.11 Jonah uses a decrementing by place value strategy that would fall in the *Additive* level of the *OGAP Subtraction Progression.* He keeps the 56 whole, decomposes the 27 by the largest place value parts, and subtracts those parts to arrive at a correct solution of 29 inches. A student who employs this strategy must have a firm understanding of place value and how to count back or subtract by those place value parts. This is another take-away strategy since the part is being removed from the whole. Some benefits of this strategy are that it is both an effective mental and written strategy, and place value is transparent throughout the process.

Figure 7.11 Jonah uses an *Additive* strategy to take away a quantity by place value parts from the whole

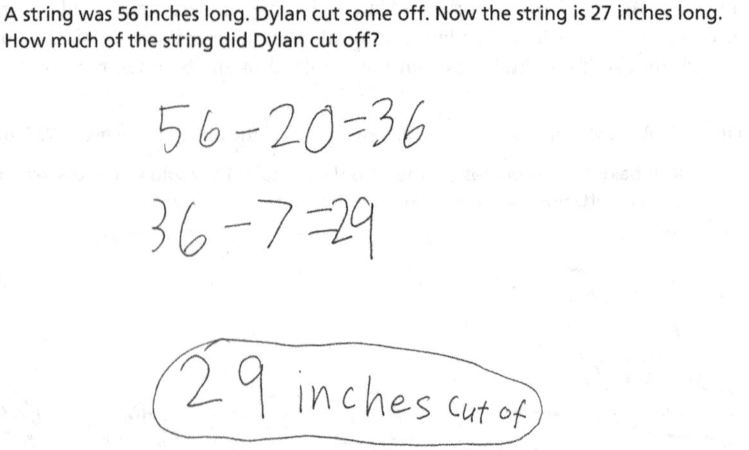

A string was 56 inches long. Dylan cut some off. Now the string is 27 inches long. How much of the string did Dylan cut off?

$$56 - 20 = 36$$

$$36 - 7 = 29$$

29 inches cut of

Find the Difference

Strategies that involve finding the difference keep both the minuend and the subtrahend whole and find the difference, or distance, between the two numbers. This strategy is a natural progression from work on a number line when subtracting. In Figure 7.12, although Camille does not draw a number line in her solution, one can almost visualize the number line in her work. She starts with 27 and adds up to get to 56, and then she adds her increments together to arrive at an answer of 29. These strategies rely on understanding of subtraction as difference. When students have at least one strategy that reflects an understanding of subtraction as difference they often have an easier time solving more challenging problems that involve comparison. Efficient use of this strategy relies on strong base-ten understanding and the ability to understand and use the relationship between addition and subtraction. Many people use this strategy when making change with money.

 Chapter 8 on Additive Situations and Problem Solving for more on problem situations that involve comparison.

Figure 7.12 Camille adds up to find the difference between 56 and 27

A string was 56 inches long. Dylan cut some off. Now the string is 27 inches long. How much of the string did Dylan cut off?

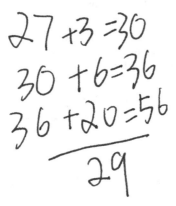

$$27 + 3 = 30$$
$$30 + 6 = 36$$
$$36 + 20 = 56$$
$$29$$

Constant Difference

The compensation strategy that works with addition does not apply to subtraction. Instead, when you add or subtract one number from one of the quantities in a subtraction problem, you need to do the same to the other quantity in order to keep the difference the same.

In Figure 7.13 Lemi changes 765 – 398 to an easier problem by adding 2 to both quantities and creating a new problem, 767 – 400, while maintaining the difference between the numbers. As you look at his work, think about why this strategy works.

Figure 7.13 Lemi uses constant difference to solve 765 – 398

Kids baked 765 cupcakes for the school bake sale. They sold some cupcakes and 398 are left. How many cupcakes did they sell?

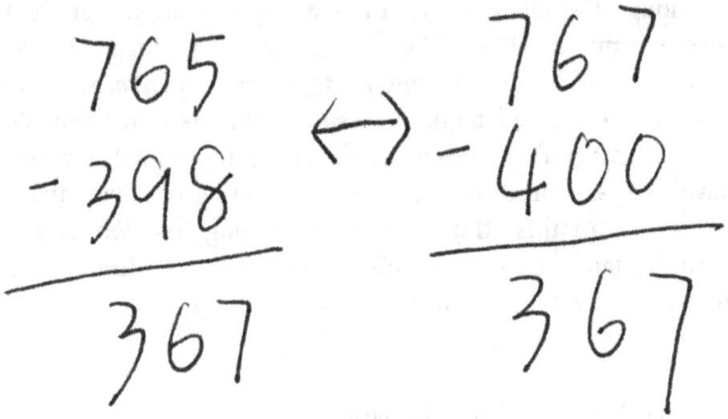

$$\begin{array}{r} 765 \\ -398 \\ \hline 367 \end{array} \longleftrightarrow \begin{array}{r} 767 \\ -400 \\ \hline 367 \end{array}$$

Figure 7.14 is an illustration of Lemi's solution to 765 – 398 on a number line. Start by locating 398 and 765 on the number line and then add 2 to both numbers. As can be seen in the illustration, the distance between the two numbers shifts but does not change in value, while the calculation has gotten significantly easier (767 – 400). The strategy relies on understanding of the relationship between the minuend and subtrahend, as well as a difference approach to subtraction.

Figure 7.14 Illustration of constant difference. Both quantities (398 and 765) have been increased by 2, but the distance or difference remains the same, resulting in an easier problem to solve

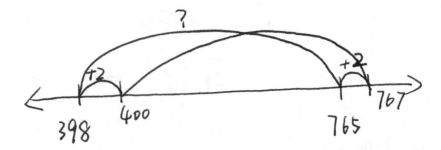

This strategy, called constant difference, *often makes sense when one of the numbers is close to a landmark or can be easily manipulated to make a simpler problem. A challenge is knowing when this strategy makes sense to use. Students who look at the numbers first to decide on the best strategy for deriving the solution are approaching calculations strategically and flexibly. Building this expectation for reasoning into instruction can be powerful for students.*

A Close Look at the Traditional US Subtraction Algorithm

As mentioned earlier, algorithms are a set of efficient steps that are carried out when solving a problem and often are taught, not student-generated. The Traditional US subtraction algorithm is the procedure most commonly taught in the US but is not the only algorithm for subtraction. Partial Differences, examined earlier in Figures 7.9 and 7.10, is also a standard algorithm for subtraction.

In Figure 7.15, Tiana uses the Traditional US algorithm to find the solution to a three-digit subtraction problem that requires regrouping across all place values. While Tiana makes a simple calculation error (15 – 8 = 8) she does execute the steps of the algorithm correctly.

Figure 7.15 Tiana uses the Traditional US algorithm for subtraction

Kids baked 765 cupcakes for the school bake sale. They sold some cupcakes and 398 are left. How many cupcakes did they sell?

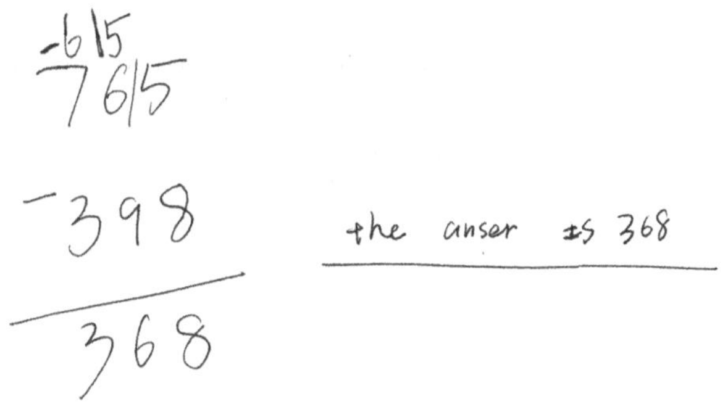

$$
\begin{array}{r}
{}^{-6}7\,{}^{1}6|{}^{1}5 \\
-\ 3\ 9\ 8 \\
\hline
3\ 6\ 8
\end{array}
$$

the anser is 368

The Traditional US subtraction algorithm is a take-away algorithm that is based on decomposition and recomposition by place value. It is important to note that while the algorithm has its foundation in place value, many students are unaware that place value plays any part in their solution as they think about the subtraction in single column exercises. While in the Partial Differences algorithm place value is maintained, or transparent, place value can be lost in the Traditional US subtraction algorithm. In order to carry out this algorithm successfully, students must decompose and then recompose the minuend (the number you start with) into parts that allow the subtrahend to be taken away by place value. The language and notation used to explain the steps increases the opaque nature of the algorithm. For example, in the problem in Figure 7.15, it is likely what Tiana said in her head as she subtracted in the hundreds place was, "6 minus 3 is 3" even though it is really 600 – 300 = 300.

Every student will likely be exposed to the Traditional US subtraction algorithm sometime in their learning experience, and because of this it is important to consider the challenges students face in using this algorithm, particularly when it has not been developed through strategies based on understanding. There is no shortage of research on the benefits, detriments, and challenges related to the Traditional US algorithm for subtraction (Baroody, 1984; Carpenter et al., 1998; Fuson & Beckmann, 2012; Fuson et al., 2005; Kamii & Joseph, 2004).

One feature that can make an algorithm more difficult occurs when there are various decisions that must be made along the way to an accurate answer (Fuson & Beckmann, 2012). When these various decisions alternate throughout the algorithm, errors are more likely. For example, in Figure 7.16 the Traditional US subtraction algorithm is used to find the answer to 426 – 188. As students move right to left finding the difference in each column, they must consider whether to "borrow" or not, carry out the task of regrouping, and then subtract. When that is complete, they move to the next column. This alternating of steps often results in more errors than if they complete all of one kind of step at a time. Consider the example in Figure 7.16. A student who completes all the steps first thinks:

- Can I take 8 from 6? No.
- Borrow from the 2, make it a 1 and make the 6 a 16.
- Can I take 8 from 1? No.
- Borrow from the 4, make it a 3 and make the 1 an 11.
- 16 – 8 is 8, 11 – 8 is 3, 3 – 1 is 2.

Figure 7.16 Using the Traditional US subtraction algorithm to solve 426 – 188

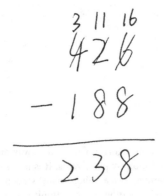

Fuson and Beckmann (2012) suggest that teaching this algorithm by completing all the regrouping, or trading, first and then going back to complete all the subtracting, makes the Traditional US subtraction algorithm easier. While the written notation for this is the same, the thought process when doing all one kind of step at a time is different.

Another challenge is that in order to complete the steps column by column, it is easy for students to abandon accurate base-ten language, perpetuating a lack of understanding about how the algorithm works and potentially compromising base-ten understanding. While instruction may begin with modeling and accurate base-ten language, it does not take long for students (and teachers) to abandon accurate language for simplicity, resulting in students forgetting what is actually happening.

Furthermore, the lack of base-ten transparency often results in students carrying out the steps in the algorithm inaccurately without considering reasonableness. One common example of this is in Figure 7.17. When 8 is too big to take from 5, the student takes 5 from 8 instead. The student carries this misunderstanding throughout the problem. The top-from-bottom subtraction error may be the result of students learning subtraction without regrouping before subtraction with regrouping. By the time they must make the decision whether to regroup or not, they have a lot of experience where that is not a consideration.

Figure 7.17 A common error in subtraction requiring regrouping

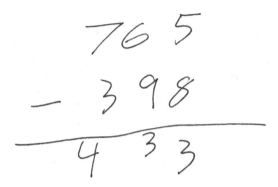

Highlighting these issues is not meant to suggest that the Traditional US algorithm itself is flawed, but rather to show that it is complex and requires thoughtful instruction. These common errors highlight the importance of focusing instruction on meaning and sensemaking when teaching students to subtract multi-digit numbers. Introducing the Traditional US algorithm after understanding is developed can avoid many of these problems. Once students have strong and flexible base-ten understanding and strategies, exploring this algorithm can be a mathematically rich rather than a procedural exercise, and students can appreciate its elegance and compactness (Bass, 2003). Ultimately students are best served when fluency with subtraction includes both efficient strategies and algorithms. As evident by examining the student work above, there are a variety of approaches based on conceptual understanding that can serve students well depending on the numbers and the situation.

Instructional Considerations for Subtraction Algorithms and Strategies

Mental math exercises can be used to encourage flexibility with subtraction. Students are often more open to think before they calculate if they are being asked to do it mentally. Additionally, when teachers encourage students to look at the numbers first in deciding the best strategy for deriving the solution, it can help students develop a habit of reasoning when carrying out operations. Consider the subtraction problem 302 – 68 posed as a mental math warm-up in a classroom. If students took a minute to think before subtracting, they might consider one of these strategies:

- Change 302 – 68 to 304 – 70 and add up to find the difference: 70 + 30 = 100; 100 + 204 + 304; 30 + 204 = 234
- Change 302 – 68 to 299 – 65 and subtract by place value to find the difference: 90 – 60 = 30; 9 – 5 = 4; 200 + 30 + 4 = 234

- $68 +$ __ $= 302$; $68 + 2 = 70$; $70 + 30 = 100$; $100 + 200 = 300$; $300 + 4 = 304$; $2 + 30 + 200 + 2 = 234$
- $300 - 60 = 240$; $2 - 8 = -6$; $240 - 6 = 234$

As teachers strategically ask students to share strategies and use visual models or representations to help all students access these strategies, students will develop flexibility and a sensemaking approach to computation.

CCSSM and Multi-Digit Subtraction

In the CCSSM, students are expected to develop and use efficient, accurate, and generalizable addition and subtraction *strategies* in grade K–2, before they are expected to use *strategies and algorithms* in grade 3 and *standard algorithms* in grade 4. As explained in the Common Core Progressions, the standard algorithms can be thought of as the "culmination" of the work done with strategies based on "place value, properties of operations, and/or the relationships between addition and subtraction" in grades K–3 (Common Core Writing Team, 2019). Throughout the standards there is a reliance on the use of visual models and base-ten understanding to make sense of addition and subtraction. Additionally, subtraction instruction is solidly anchored to developing and using the relationship between addition and subtraction.

 Chapter 6 for a more thorough understanding of addition and subtraction in the CCSSM and instructional considerations when developing procedural fluency with multi-digit addition and subtraction.

Summary

This chapter focused on developing strategies and algorithms for subtraction. An important point is that base-ten understanding and fluency with subtraction go hand in hand and can be developed simultaneously. Inventing strategies for adding and subtracting multi-digit numbers and developing strong base-ten understanding are mutually reinforcing (Carpenter et al., 1998; Fuson & Briars, 1990).

- Visual models, place value, properties of operations, and the relationship between addition and subtraction all play an important role in the development of strategies and algorithms for subtraction.
- When students look at the numbers first to decide on the best strategy for deriving the solution, they are approaching calculations strategically and flexibly. This can help to develop a reasoning approach to operations. Mental math exercises can often be used to encourage flexibility with computation.
- There are two distinct meanings of subtraction: take-away and difference. These two meanings influence the way students think about subtraction and the strategies and algorithms they may employ to solve subtraction problems.
- Students should have both efficient strategies and algorithms for subtraction towards developing procedural fluency.
- There is a danger in introducing a standard algorithm before developing understanding through strategies based on understanding. Students who are reliant on a procedure without understanding are at a disadvantage, as the procedure is subject to deterioration, and errors are more likely.

Looking Back

1. **Four Categories of Subtraction Strategies:** This chapter discusses four general categories of subtraction strategies: decompose and recompose by place value, decrement by place value, find the difference, and constant difference. Analyze the student work in Figure 7.18 and 7.19 to:

    (a) Describe the strategy the student used to solve the problem.
    (b) Determine what category best describes the student's strategy.

Figure 7.18 James' and Siree's solutions to the fundraiser problem

> Ms. Brown's class collected $72 for the school fundraiser, and Mr. Smith's class collected $29 for the fundraiser. How much less money did Mr. Smith's class collect than Ms. Brown's?

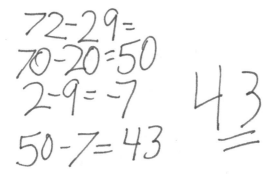

James' solution

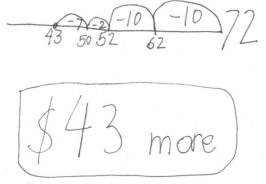

Siree's solution

Figure 7.19 Stephen's and Logan's solutions to the park problem

There were 134 kids playing at the park. Then 59 kids left to have lunch. How many kids were still playing?

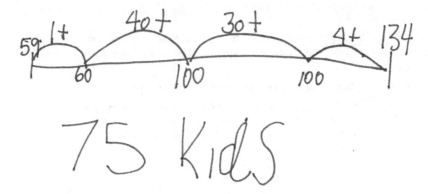

Stephen's solution

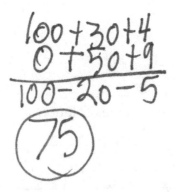

Logan's solution

2. **Two Meanings of Subtraction:** It is important for students to understand subtraction as both taking away and distance. Reconsider the student work in Figures 7.18 and 7.19. Based on the strategy used to solve the problem, is the student thinking about subtraction as take-away or as difference/distance?

3. **Using the OGAP Subtraction Progression:** With increased understanding of properties of addition, number relationships, and our base-ten number system, students can use more efficient strategies for subtraction like those reflected at the top of the progression. Compare Heather's response in Figure 7.20 to Camille's response from earlier in the chapter.

(a) What is similar about the understanding both students demonstrate?

(b) Where does each response fall on the progression? What accounts for the difference?

Figure 7.20 Heather's and Camille's solutions to the string problem

A string was 56 inches long. Dylan cut some off. Now the string in 27 inches long. How much of the string did Dylan cut off?

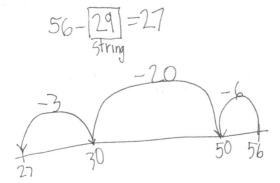

Heather's solution

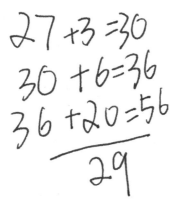

Camille's solution

4. **Visual Models for Subtraction:** A valuable instructional approach involves helping students make connections between models and strategies.

(a) Create a list of questions that could be asked to help students make connections between the visual models in Micah's and Rico's strategies in Figure 7.5.

(b) Consider Rico's representation of base-ten blocks. What are some challenges associated with using base-ten models for subtraction?

(c) Is Rico able to successfully overcome the challenges? What evidence in his work supports this analysis?

5. **Comparing Two Student Solutions:** Successfully executing efficient subtraction strategies with multi-digit numbers involves flexible use of addition properties, number relationships, and base-ten fluency. How is the understanding demonstrated in Jimmy's work (Figure 7.21) similar to the understanding demonstrated in Mason's work from earlier in the chapter? How are they different?

Figure 7.21 Mason's and Jimmy's solutions to the string problem

A string was 56 inches long. Dylan cut some off. Now the string in 27 inches long. How much of the string did Dylan cut off?

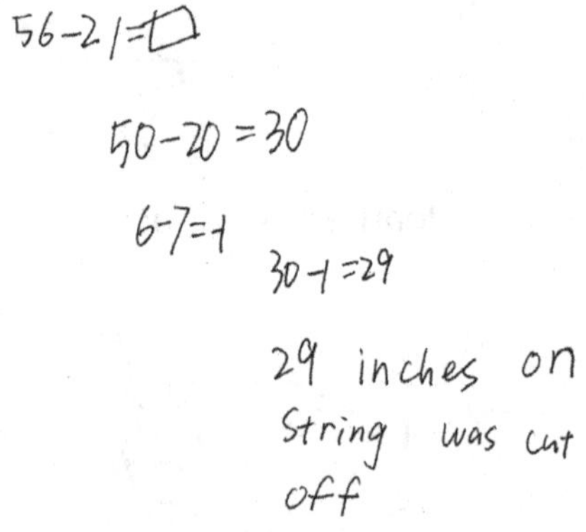

$$56 - 21 = \square$$

$$50 - 20 = 30$$

$$6 - 7 = 1$$
$$30 - 1 = 29$$

29 inches on
String was cut
off

Mason's solution

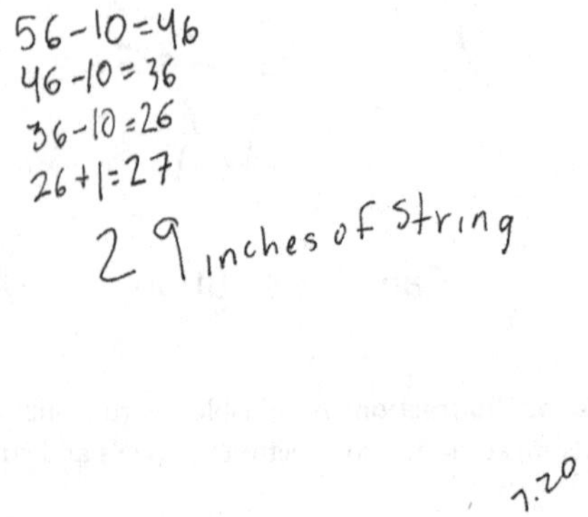

$$56 - 10 = 46$$
$$46 - 10 = 36$$
$$36 - 10 = 26$$
$$26 + 1 = 27$$

29 inches of String

7.20

Jimmy's solution

Instructional Link

Use the information presented in this chapter to consider the ways your math program and instruction offer students a research-based approach to gaining fluency and understanding of subtraction.

1. Do your instructional materials provide opportunities for students to:
   (a) Experience a variety of subtraction strategies and make connections between strategies?
   (b) Explore the concepts of addition and subtraction in tandem?
   (c) Use visual models to make sense of subtraction concepts and solve subtraction problems?

2. To what degree do your math curriculum materials align with developmentally appropriate exposure to strategies and algorithms for subtraction for the grade you teach?
3. To what degree do you collect and analyze evidence in student solutions to inform instruction and student learning?
4. Based on your previous answers, what modifications might be considered to help your students gain a deeper conceptual understanding of subtraction?

8
Additive Situations and Problem Solving

Big Ideas

- The structure of additive problems influences both the strategy that students will use to solve the problem and the difficulty of the problem.
- There are twelve different additive situations that vary based on the action or relationships in the problem and the quantity that is unknown.
- The nature of the quantities, units, and language in the problem also influence the level of difficulty for students.
- It is important for students to have experience solving a variety of problem contexts and situations in the primary grades.
- Students can learn to be successful problem solvers by focusing on making sense of the problem context, situation, and quantities before deciding on a strategy to use.

The ability to solve problems that reflect a variety of mathematical situations is an important component of mathematical proficiency. This chapter focuses on the importance of problem structures and the different additive problem situations that are produced by varying the action or relationships in the problem and the quantity that is unknown. The level of challenge or difficulty of a problem is influenced by these problem structures as well as by the nature of the quantities, units, and language in the problem. With this knowledge, teachers can strategically select problems, adjust the level of difficulty for diverse learners, and ensure that students solve a range of different types of problems to further the development of additive reasoning.

We begin with a vignette of a second-grade teacher, Mr. Paul. At the beginning of the year, Mr. Paul gave his students the OGAP pre-assessment which included the following two related problems:

Kai had 23 stickers. He gave 16 stickers to Jaden.
How many stickers does Kai have left?

Sam had 16 marbles. Nadia gave him some more marbles.
Now Sam has 23 marbles. How many marbles did Nadia give Sam?

Although Mr. Paul knew that both problems could be solved with subtraction (23 – 16), he was surprised to see that most students responded to the two problems differently. He noticed the following three patterns in his student work:

- Some students solved both problems correctly but used different strategies.
- Some students who could solve the first problem successfully were not successful in solving the second problem.
- Most students did not recognize a relationship between the two problems.

Figure 8.1 shows an example of a student who solved both problems correctly but used different strategies. Nadia solved the first problem by using the Traditional US algorithm for subtraction, a strategy at the *Additive* level of the *OGAP Subtraction Progression*. However, for the second problem she used a lower level *Counting* strategy; she counted on from 16 to 23 by ones, using the number line to keep track of the 7 jumps.

Figure 8.1 Nadia uses strategies on different levels of the *OGAP Subtraction Progression* to solve related problems

Kai had 23 stickers. He gave 16 stickers to Jaden. How many stickers does Kay have left?

Sam had 16 marbles. Nadia gave him some more marbles. Now Sam has 23 marbles. How many marbles did Nadia give Sam?

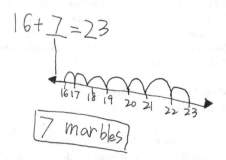

Mr. Paul also noticed that several students who could solve the first problem success-fully were not successful in solving the second one. As shown in Figure 8.2, Terrence solved both problems by direct modeling and counting all by ones. He solved the first problem by drawing out 23 circles, crossing out 16, and counting up the remain-ing circles by ones to get the correct answer of 7. However, for the second problem, he drew out 16 and 23 circles and counted them all to get a total of 39.

Figure 8.2 Terrence uses direct modeling and counting to solve both problems but uses the wrong operation for the second problem

Kai had 23 stickers. He gave 16 stickers to Jaden. How many stickers does Kay have left?

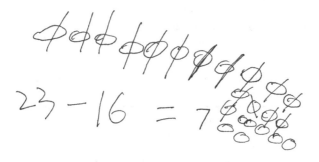

Sam had 16 marbles. Nadia gave him some more marbles. Now Sam has 23 mar-bles. How many marbles did Nadia give Sam?

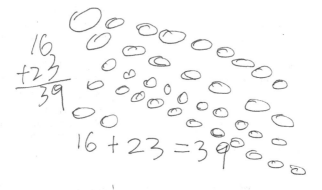

Finally, Mr. Paul noted that only one student seemed to have noticed the relation-ship between the problems. Fatima counted back from 23 by ones to solve the first problem correctly. For the second problem, she then wrote "from the problem I just did, I only needed to flip the equation 16 + 7 = 23 around. Nadia gave Sam 7 marbles." Mr. Paul thought it was interesting that even though she recognized the relationship between the problems, Fatima was thinking of the first problem as a subtraction problem and the second one as an addition problem.

The patterns Mr. Paul noted reflect research that shows that children solve word problems by modeling the action or relationship in the problem (Carpenter, Fennema, Franke, Levi, & Empson, 2015). In the first problem the stickers are being given away and the problem involves a separating or taking away action. In the second problem, the marbles are being given to a set that is already there, so the action is joining or adding to. Therefore, children will often model this as an addition problem with a missing addend rather than as a subtraction problem. Furthermore, as will be explained in the following section, research shows that this second problem is generally more difficult for students to solve than the first one.

Problem Types and Difficulty

Researchers have identified four main types of actions or relationships in addition and subtraction problems as shown in Table 8.1: (1) add to (joining), (2) take from (separating), (3) put together/take apart, and (4) compare. Review the descriptions in Table 8.1 to compare these problem situations.

Table 8.1 Four types of additive situations based on the action or relationship in the problem

Additive Situation	Action or Relationships	Temporal Nature	Example
Add To	Joining, increasing, or adding to a given set	Action over time	There are 3 children on the playground. Then 2 more children come to the playground. How many children are on the playground now?
Take From	Removing, separating from, or decreasing from a given set	Action over time	There are 5 children on the playground. Then 2 children leave the playground to go home. How many children are on the playground now?
Put Together/ Take Apart	Putting together or combining parts into a whole	Static relationship	There are 3 children and 2 adults on the playground. How many people are on the playground?
Compare	Comparison of two distinct sets, with a difference	Static relationship	There are 5 children and 3 adults on the playground. How many more children are there than adults?

Problems that involve adding to or taking from involve actions that occur over time, while problems that involve putting together or taking apart or comparison do not; instead these situations involve relationships between the quantities. As shown on the *OGAP Additive Framework*, at first, young students understand the structure of a problem by creating either a mental image, a drawing, or a concrete model of what is happening, making problems with an action more accessible to students than problems with a static relationship, or no obvious action happening over time.

Add To and Take From Problems

Add To problems that involve combining or joining are the most accessible because children can visualize, draw out, or build both quantities with objects and then collect them to create the total quantity. Take From problems that involve separating or taking away are a little more challenging because one quantity is part of the other. With physical models, once the part is separated or taken away the initial quantity is no longer there, making it hard to see the relationship.

Consider Noah's solution to a Take From problem shown in Figure 8.3. Noah solves the problem by drawing and counting out 8 pennies. He then crossed out four pennies to represent the pennies given away and sees that there are 4 left. The drawing helps to show that both quantities of four (the 4 pennies given away and the 4 that remain) are part of the original 8. With physical models, this part–whole relationship would be harder to see, as the counters would be gone.

Figure 8.3 Noah's *Early Counting* strategy models a Take From problem

Sara had 8 pennies. She gave 4 pennies to Samuel. How many pennies did she have left?

Put Together/Take Apart Problems

Put Together/Take Apart problems are similar to Add To and Take From problems but the fact that there is no action over time makes them somewhat more difficult for young students. These problems involve a set of something and its subsets, for example: a box of crayons that are red and blue, a group of pets that are dogs and cats, or a plate of vegetables (peas, carrots, and onions). Understanding a Put Together/Take Apart situation involves understanding class inclusion—e.g., dogs and cats are a subset of pets—which adds a layer of complexity. However, they also provide context for students to develop a conceptual understanding of the part–whole relationship as well as the commutative property because it is easier to see that the order of the addends does not matter when there is no action over time. In Figure 8.4, Lindsay's solution shows 5 + 4 = 9 even though the problem lists the quantities in a different order. The part–whole relationship and commutativity can both emerge from the models students create for this type of situation.

Figure 8.4 Lindsay draws a model for a Put Together/Take Apart problem that illustrates part-whole and commutativity

There were 4 red flowers and 5 purple flowers in Jack's garden. How many flowers were in Jack's garden?

Compare Problems

Problems that involve comparison offer another element of complexity that makes them more difficult or challenging for students to solve. In comparison situations, the difference is a relationship between two quantities rather than a quantity that is physically present. These problems are also linguistically challenging because phrases like "seven more than" or "15 fewer than" communicate two important pieces of information—a quantity (15) and a relationship (fewer than). Compare Noah's solution to a Compare problem in Figure 8.5 to his solution to the Take From problem in Figure 8.3.

Figure 8.5 Noah's *Early Counting* strategy for a Compare problem

There were 6 children playing basketball. There are 4 children on the swings. How many more children are playing basketball than are on the swings?

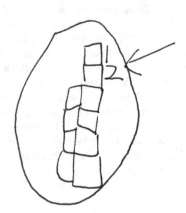

To solve this Compare problem, Noah does not put out 6 blocks and take away 4 as he did to solve the Take From problem in Figure 8.3. Instead, he counts out and draws 6 blocks to represent the children playing basketball and 4 blocks to represent the children on the swings. He then lines them up next to each other and matches up the blocks in each tower to see that there are 2 more in the group of 6. The difference between the Take From situation in the problem in Figure 8.3 (giving away pennies) and the Compare situation (how many more) in the problem in Figure 8.5 led to a different *Early Counting* strategy.

Variations Based on the Location of the Unknown

Problems within each of the four classes of problem situations can also be varied by changing the quantity that is unknown. For example, consider the Add To problem from Figure 8.1.

> There are 3 children on the playground. Then 2 more children come to the playground. How many children are on the playground now?

This same problem situation can be presented with the *change*, or amount added, unknown:

> There are 3 children on the playground. Then some more children come to the playground. Now there are 5 children on the playground. How many children came to the playground?

It can also be constructed to have the *start*, or beginning amount unknown:

> There were some children on the playground. Then 2 more children come to the playground. Now there are 5 children are on the playground. How many children were on the playground to start?

Think about how children's strategies for modeling these variations with concrete objects or drawings might differ.

The problem shown in Figure 8.6 reflects a Take From action: there is a set of balloons to start with and some fly away. However, in this problem it is the

Figure 8.6 Mandy's *Early Counting* solution models the situation with drawings

> Sarah had a bunch of 14 balloons. She let some go and they flew away. She still has 5 balloons in her hand. How many balloons flew away?

amount that flew away, or the change, that is unknown while the amount that is left (5 balloons) is known. Consider how Mandy solves this problem in Figure 8.6. At first, Mandy's solution looks like Noah's solution in Figure 8.3. Mandy first draws and numbers 14 balloons and then crosses some out. However, while Noah knew how many to cross out, Mandy must cross out an unknown number of balloons until she has 5 left. Because she has numbered the original set and takes away from one end, she knows when she has reached 5. She then has to go back and count how many she crossed out to obtain the answer of 9. Although this may seem like a subtle difference, it has implications for both the strategies children use and difficulty of the problems. In Figure 8.7, Louisa uses a slightly more advanced counting back strategy. She does not need to model the 14 and the number line helps her keep track of the 9 jumps back to 5.

Figure 8.7 Louisa uses a *Counting* strategy to count back the distance from 14 to 5 on a number line

Sarah had a bunch of 14 balloons. She let some go and they flew away. She still has 5 balloons in her hand. How many balloons flew away?

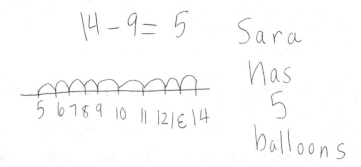

When the location of the unknown is considered, there are 12 different additive situations, as shown in Table 8.2.

For some problem situations, an equation or number sentence that models the situation can be directly derived from the problem situation. Consider the Take From with the change unknown problem in Table 8.2:

Jon had 8 crayons. He gave some away and now he has 5. How many crayons did he give away?

This situation can be translated into the equation $8 - x = 5$. However, to solve the problem, children may either add $5 + 3 = 8$ or subtract $8 - 5 = 3$. Therefore, while the equations listed for each of the problem situations in Table 8.2 offer ways to model each problem situation, they are not always a model *of* the solution. Another way to think about this is that the equation reflects the situation but does not

Table 8.2 Twelve additive situations for the combination 5 + 3 = 8

	Result Unknown	Change Unknown	Start Unknown
Add To	Jon had 5 crayons. He got 3 more crayons. How many crayons does he have now? 5 + 3 = ?	Jon had 5 crayons. Then he got some more crayons and now he has 8. How many more crayons did he get? 5 + ? = 8	Jon had some crayons. He got 3 more crayons and now he has 8. How many crayons did he start with? ? + 3 = 8
Take From	Jon had 8 crayons. He gave 3 away. How many crayons does he have now? 8 – 3 = ?	Jon had 8 crayons. He gave some away and now he has 5. How many crayons did he give away? 8 – ? = 5	Jon had some crayons. He gave 3 away and now he has 5. How many crayons did he give away? ? – 3 = 5
	Total Unknown	**Both Parts Unknown**	**Part Unknown***
Put Together/ Take Apart	Jon has a box of 5 blue crayons and 3 red crayons. How many crayons are in the box? 5 + 3 = ?	Jon has a box of 8 blue and red crayons. How many blue crayons and red crayons could be in the box? ? + ? = 8	Jon has a box of 8 crayons. 5 of them are blue and the rest are red. How many red crayons are in the box? 5 + ? = 8 8 – 5 = ?
	Difference Unknown	**Bigger Unknown**	**Smaller Unknown**
Compare	Jon has 8 crayons and Robert has 3 crayons. How many more crayons does Jon have than Robert? 3 + ? = 8 8 – 3 = ?	Robert has 3 crayons. Jon has 5 more crayons than Robert has. How many crayons does Jon have? 3 + 5 = ?	Jon has 8 crayons. He has 5 more crayons than Robert has. How many crayons does Robert have? 8 – 5 = ? ? + 5 = 8

* Note there are two versions because either addend can be unknown
Adapted from NGA & CCSSO, 2010; NRC, 2009.

always provide directions for how to calculate the answer. This is why it is important for students to make sense of the problem situation before deciding how to compute.

Add To, Take From, and Put Together/Take Apart problems all offer the opportunity for students to understand the inverse relationship between addition and subtraction because the action or situation can be reversed: In the examples in Table 8.2, the action of giving away crayons undoes the action of getting crayons, and taking apart the two colors of crayons undoes the putting together in one group. These problem situations are useful for highlighting the inverse relationship between these operations, an important foundation for both additive reasoning and fact fluency.

Put Together/Take Apart problems with both addends unknown offer a particularly rich opportunity for students to develop part-whole understanding and the relationship between addition and subtraction, critical foundations of additive reasoning. Figure 8.8 shows a student response to this type of problem where the sum is 10. Lucy's response shows that she could use a drawing of 10 pumpkins to find one solution and then draw her own solutions from there. Her solutions offer a great opportunity to discuss the patterns in the combinations as well as the concept of compensation (adding one to one addend and subtracting one from the other addend does not change the total).

Figure 8.8 Lucy's response. Lucy shows evidence of using a pattern to find multiple combinations of 10

Ms. Cho had 10 pumpkins. She decorated some pumpkins by carving faces. What combinations could she have of decorated and plain pumpkins? Find as many possible combinations as you can.

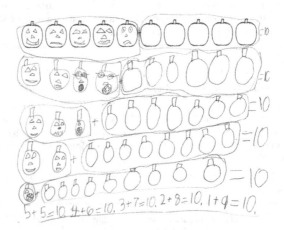

Knowing how to decompose 10 into parts is also very important for developing efficient strategies for addition and subtraction that use base-ten understanding.

 Chapter 6 for more on compensation and addition strategies and Chapter 7 for subtraction strategies.

Additional Considerations: Language, Context, Quantities, and Units

It is also important to be aware of some other ways in which problem structures can vary and impact students' sensemaking, the strategies they use, and difficulty of the problem. In this section we discuss the impact of language, problem context, and the nature of the quantities and units in the problem. This information, along with knowledge of students' developing understanding, should be used when selecting problems for students to solve.

Language

There are additional variations in Compare problems, depending on whether the comparison is phrased as a version of "more than" or "less than." Consider the following variations of a Compare problem where the difference is unknown. Which of the following problems do you think would be easiest for children to solve?

(a) Ali has 9 balloons. Lisa has 13 balloons. How many more balloons does Lisa have than Ali?

(b) Ali has 9 balloons. Lisa has 13 balloons. How many balloons does Ali need to get to have the same amount as Lisa?

(c) Ali has 9 balloons. Lisa has 13 balloons. How many fewer balloons does Ali have than Lisa?

All three of these problems involve comparing the two quantities of 9 and 13 and can be solved by adding up from 9 to 13 or subtracting 9 from 13. However, version (a) has "more than" language that may be more familiar and accessible for students than the "fewer than" language in version (c). The language in version (b) is even more accessible because it turns the static nature of the comparison into an active one of equalizing, making it closer to an Add To problem (Fuson, Carroll, & Landis, 1996).

Context

The context of a problem is another important consideration. Consider the following two problems. What is similar and what is different about the two problems?

- Tanya mowed the lawn for 20 minutes, stopped to refill the mower with gas, and then mowed the lawn some more. She mowed the lawn for 47 minutes altogether. How long did Tanya mow the lawn after she refilled the mower with gas?
- Kesha played for 20 minutes, then she stopped to eat lunch. After lunch she played some more. Kesha played for 47 minutes altogether. How long did Kesha play after lunch?

Both problems are Add To situations with the change unknown involving minutes and both can be modeled with the equation $20 + \_\_\_ = 47$. However, the first problem involves a context, mowing a lawn with a gas-powered lawn mower, that may not be familiar to students who don't live in a single-family home with a lawn. The second problem involves situations (playing and lunch) that are more likely to be familiar for all students.

Quantities

The magnitude and nature of the numbers or quantities in a problem also influence difficulty, as well as the relationship between the quantities. Increasing the magnitude of the quantities will add more challenge to a given problem. When selecting and creating problems, teachers should also consider how the relationship between the numbers in the problem will affect the level of access or challenge for students. For example, number combinations are that are doubles or near doubles may make a problem easier to solve, and combinations that involve a computation that crosses over one or more place value units will be more challenging. For subtraction, the distance between numbers (how close or far apart they are from each other) may also affect the difficulty of the problem.

Type of Units

Finally, consider the following two problems. As you read the problems, think about how students might visualize or draw the problems differently.

- Tre had 46 Yankees cards in his baseball card collection, and the rest of his cards are Mets cards. He has 100 cards in his collection. How many Mets cards does he have?
- Juan drove from Philadelphia to New York. He left Philadelphia, drove 53 miles, and then stopped to eat lunch. After lunch, he drove the rest of the way to New York. He drove 97 miles altogether. How far did Juan drive after lunch?

You may have noticed that the first problem is about baseball cards, a discrete quantity that can be counted or visualized as a collection. The second problem is about distance, and the units are miles, a continuous quantity that is measured. The second problem also lends itself more directly to a number line model, as shown in Drew's solution in Figure 8.9.

Figure 8.9 A problem involving distance can be represented on a number line

Juan drove from Philadelphia to New York. He left Philadelphia, drove 53 miles, and then stopped to eat lunch. After lunch, he drove the rest of the way to New York. He drove 97 miles altogether. How far did Juan drive after lunch?

Drew solves this problem by locating 53 and 97 on the number line and then finding the distance between those two points. The fact that the problem situation involves distance makes this a natural choice and may have influenced his choice of a model. In this way, teachers can be intentional about varying problem structures to introduce or emphasize different models and strategies for solving additive problems.

CCSSM Expectations on Problem Situations

The CCSSM expects students to be able to solve all 12 types of additive problem situations by the end of second grade. However, the expectations for each grade level reflect the research findings on the relationship between problem structures and difficulty. Table 8.3 shows the new problem structures that students are

Table 8.3 Summary of problem situations expected at each grade level in the CCSSM

Grade	New Problem Situations	Position of Unknown
Kindergarten	Add To	Result unknown
	Take From	Result unknown
	Put Together/Take Apart	Total and Both Addends unknown.
Grade 1	Compare	Difference, Bigger and Smaller unknown with "more" language
	Add To and Take From	Change unknown
	Put Together/Take Apart	Addend unknown
Grade 2	Compare	Bigger and Smaller unknown with "fewer" language
	Add To and Take From	Start unknown
	Multi-step and multi-situation	Easier versions (e.g., result or total unknown)

expected to be able to solve at each grade level. Note that this represents a cumulatively growing set of problem situations over time.

In kindergarten, the problem situations are easily modeled with concrete objects or drawings. Of note is the expectation that kindergarten students solve Put Together/Take Apart problems with both addends unknown. As explained in the progression for *Operations and Algebraic Thinking* (Common Core Writing Team, 2019), these problems "play an important role in Kindergarten because they allow students to explore various compositions that make each number" (p. 19).

In grade 2, children are expected to solve two-step problems that may involve multiple situations. An example of a multi-situation problem is shown in Figure 8.10. The problem involves a Put Together/Take Apart with the result unknown (a set of red and blue buttons) and a Take From with result unknown (giving away buttons).

Figure 8.10 Marla's solution to a two-step multi-situation problem

Janelle had 76 red buttons and 45 blue buttons. Then she gave 25 of her buttons to her friend. How many buttons does Janelle have now?

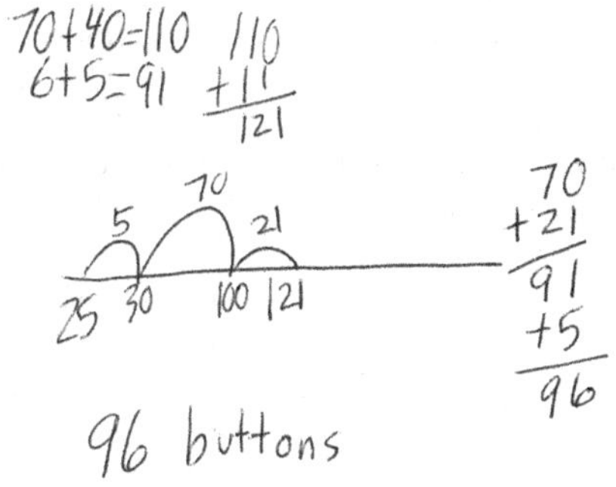

As Marla's solution shows, this problem cannot be represented with a single equation or computation, but rather has two separate parts. Marla uses an *Additive* strategy involving decomposing and recomposing by tens for the addition of 76 and 45 and a slightly less sophisticated *Transitional* strategy to find the difference between 121 and 25 on a number line.

While the standards clarify which problem situations students should be successfully able to solve at each grade level, this does not mean that students should only work on those specific problem types. At every grade level, more or less complex problem situations or structures may be used to provide students with opportunities for challenge or access.

Instructional Implications

Problem structure influences the strategies students use to solve problems as well as problem difficulty. Although it is not important for students to know the different names or types of additive situations, this is important information for teachers. One way this information can be useful is to ensure students are being asked to solve a variety of problem types in relation to grade level expectations. In OGAP training, when teachers are asked to create word problems for different addition and subtraction equations, they almost universally construct Add To and Take From problems, and sometimes Put Together/Take Apart problems. They almost never think about problems with a Compare situation. Teachers and curriculum materials may be consciously or unconsciously neglecting these types of problems when assigning work for students.

Knowledge of problem structures can also support teachers in differentiating and targeting instruction for students based on developing understanding. Problems can be engineered to be more accessible or more challenging by changing the numbers, the language, or the problem situation (e.g., replacing a change unknown problem with a result unknown problem).

The front page of the *OGAP Additive Framework* includes information on *Problem Contexts* and *Problem Structures*. The chart of Additive Situations is shaded to show the CCSSM grade level expectations and problem difficulty. This information can be used to inform instructional decisions and ensure that students interact with a range of grade-level appropriate problems. For example, if students are using strategies at the *Additive* level to solve given Add To and Take From problems with the result unknown, a teacher can provide more challenge by increasing the magnitude of the quantities in the problem, adapt the problems so that the change or start is unknown, or select problems that involve comparison.

Look at the arrow on the left side of all three progressions in the *OGAP Additive Framework*, which states: "The strategies students use move back and forth across the levels as they learn new concepts and/or interact with new problem structures and situations." This arrow illustrates one of the patterns that Mr. Paul noticed: The strategy a student uses may depend on the structure of the problem. When the problem structure is more difficult or unfamiliar, students often use lower level strategies that more directly model the action or relationship in the problem. This also means that teachers can help students learn to solve more difficult problem structures by focusing on modeling what is happening in the problem. In the next section, we describe one strategy that has proven useful for supporting student sense-making around word problems.

Instructional Strategies for Problem Solving

Many people have been taught that word problems have clues or "key words" that can unlock the operation needed to solve them. Unfortunately, in reality, most problems do not contain such clear directions. For example, consider the following problems and see if you can find a key word that signals the correct operation to use to solve it:

- Nine birds are in a tree. Four of these birds are blue. The other birds are red. How many birds are red?
- Siling collected 8 seashells which was 2 times more than what Jon collected. How many seashells did Jon collect?
- How many legs do 6 elephants have?
- Lester has 3 fewer oranges than Jacob. Lester has 8 oranges. How many oranges does Lester have?

As these examples show, many problems do not contain key words, or the words they contain do not signify the appropriate operation for solving the problem. Teaching children that words or phrases like *altogether* and *in all* mean addition, or *left* and *fewer* mean subtraction, does not set students up for success in the long term. Imagine what happens when a well-intentioned primary grade teacher tells students that the word altogether signifies addition. For many of the problems that students see in those grades, this may be true. But then in third grade, they may see a problem like this:

Sam had 3 boxes with 4 balloons in each. He bought twice as many balloons at the store. How many balloons does he have altogether?

As this example shows, the keyword strategy will not work with most multi-step problems and more complex problems. In fact, research shows that a key word strategy is not effective and can in fact, be detrimental (Clement & Bernhard, 2005; Drake & Barlow, 2007; Sowder, 1988). One reason is that by focusing students on key words, we are teaching them to ignore the surrounding context, in effect preventing the sensemaking that we know is important for problem solving.

In the example shown in Figure 8.11, Isabel produces a non-sensical answer by attempting to subtract 83 from 79. Although we do not know for sure, it may be that she focused on the words "shorter than" as a clue to subtract and pulled out the numbers in the order they were presented in the problem.

Figure 8.11 Isabel's work shows a *Non-Additive* strategy that does not model the problem situation

The red ribbon is 79 inches long. The red ribbon is 83 inches shorter than the yellow ribbon. How long is the yellow ribbon?

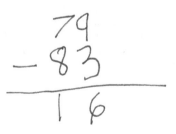

Fortunately, there are effective alternatives to the key word strategy. Math educators and researchers agree that a focus on helping students make sense of the problem and the quantities should be a priority. The first CCSS Standard for Mathematical Practice is "Make sense of problems and persevere in solving them." When students truly understand the problem, they are more likely to choose an appropriate strategy to solve it. A research study by Hegarty, Mayer, and Monk (1995) showed that less successful problem solvers focused on words and numbers while the successful problem solvers constructed a mental model of the problem situation.

Making sense of word problems involves engaging with text, and making sense of text involves synthesizing, finding important information, and drawing inferences (Morrow, 1985). In the next section we present one proven strategy for helping students make sense of word problems that draws on these ideas from research-based literacy and reading comprehension strategies (Gambrell, Koskinen, & Kapinus, 1991; Metsisto, 2005).

A Literacy Strategy to Engage Students in Making Sense of Word Problems

The strategy involves four parts:

1. Remove the question from the problem and have students read and retell the situation (read and retell).
2. Have students identify the important information in the problem (analyze).
3. Have students generate questions that can be asked and answered given the problem situation (predict).
4. Have students solve the original problem and/or the problems that were generated (solve).

This sequence is illustrated below with the following multi-step word problem. However, it can be done with any word problem.

Jane is making a fruit salad for a picnic. She went to the store and bought 8 apples, 5 oranges, 6 pears, and a bunch of bananas. She has 25 pieces of fruit altogether. How many bananas did she buy?

Read and Retell. The first step is to remove the question from the problem and read or have students read the problem situation. Many educators have found that engaging in multiple reads as well as having students read, hear, and see the problem can be helpful.

Now have students turn to a partner and paraphrase the story in their own words. After one to two minutes, ask a student to retell the story to the class. This is a good opportunity to check in on student understanding of the situation and clarify any vocabulary or aspects of the context that may be unfamiliar (e.g., What is a fruit salad? What does it mean to have a bunch of bananas?).

Analyze. The next step is to ask students to name the important information in the story. As students offer important facts, list them on chart paper or on the board. For example, students might suggest:

- Fruit salad—mixture of fruits
- 8 apples

- 5 oranges
- 6 pears
- 1 bunch of bananas
- 25 pieces of fruit
- Fruit: apples, oranges, pears, bananas

This step is important because it engages students in quantitative analysis or focusing on the meaning of the quantities in the situation, rather than just the values. (Thompson, 1994 as cited in Clement & Bernhard, 2005). A focus on the values only leads to what many refer to as number grabbing: pulling out the numbers from the problem and then inserting an operation rather than making sense of the quantities in relation to the situation. This is also important because not all quantities that are involved are listed in the problem (e.g., the number of bananas in the bunch), and sometimes quantities are included that are not needed to solve the problem.

Predict. Next have students work with a partner to generate at least two questions that can be asked and answered given the context and information provided in the problem. Some possibilities they may come up with include addition (How many apples and oranges?), comparison (How many more apples than oranges?), or even fractions (What fraction of the fruit are the oranges?).

After partners have completed their list of questions have students share their questions with the class. Post all the questions for the full class to see. Students are usually very surprised by all the different questions that can be answered using similar information.

Solve. The final step is to reveal the original problem and have students solve it and/or the problems they generated. This is also an opportunity for differentiation since different problems have been created. As students work on solving, encourage them to visualize, draw a picture or diagram, use concrete materials, or even act out the problem. The goal is for students to construct a mental model of the problem situation before deciding which operations to use.

This word problem strategy should be applied multiple times for students to reap the benefits. Whenever students run in to difficulty solving problems on their own, they can be refocused by prompting them to cover the question and retell the problem situation. Another useful variation of this strategy is to give students problems without numbers. The most important thing is to focus students on making sense of the situation before they choose an operation or start computing a result.

Summary

This chapter focused on problem structures for addition and subtraction.

- Teachers should be familiar with the 4 different additive problem situations and 12 variations that students are expected to be able to solve by the end of second grade.
- The strategies students use to solve additive problems are dependent on their developing understanding and the problem structures (e.g., number size and situation).
- Problems without action and with an unknown in the starting or change position are more difficult for students to solve because they are more

difficult to model. This information can be used to inform instruction on a regular basis.

- Supporting students to solve word problems involves helping them make sense of problem situations, focusing on the meaning of the quantities, and developing a mental model of the situation before choosing an operation.

Looking Back

1. **Analyzing Additive Problem Situations:** The structures of additive problem situations can influence the strategies students use to solve the problem. Read through the following additive problem situations and use Table 8.2 to classify them by problem type.

Problem 1

Nicholas had 891 pennies.
He donated 499 of his pennies to his class penny collection.
How many pennies does he have left?

Problem 2

2 young kids were on the swings.
Then some older kids got on the swings.
Now there are 5 kids on the swings.
How many older kids are on the swings?

Problem 3

There are 22 students in Paul's class.
There are 15 students in Harry's class.
How many fewer students are in Harry's class than Paul's class?

Problem 4

Min has 5 balloons.
3 balloons are blue.
The rest are yellow.
How many balloons are yellow?

Problem 5

Maya had 9 crayons and then she got 6 more.
Now Maya has 5 more crayons than Manny.
How many crayons does Manny have?

Problem 6

A stuffed bear is 14 inches tall.
A stuffed giraffe is 7 inches taller than the stuffed bear.
How tall is the stuffed giraffe?

(a) For each problem, identify the important problem structures including additive situation, number complexity, language, and units.
(b) How do these structures impact the level of difficulty?

(c) For problems 2, 4, and 5, how do you anticipate students would model the problem with concrete material or drawings?

2. **A Classroom Vignette:** Ms. Smith used the following task as an exit ticket after a lesson in a unit on addition. As she analyzed the student responses she noticed several students were not successful with the task even though they were able to solve problems with numbers of the same magnitude previously. The student work shown in Figure 8.12 is representative of a common challenge some students had with the task.

    (a) Use Table 8.2 to categorize the problem.
    (b) Describe issues or errors evidenced in the student work.

Figure 8.12 A representative sample of the incorrect student responses in Ms. Smith's class

The express bus takes 126 minutes to go from Philadelphia to New York.
The local bus takes 115 minutes longer to go from Philadelphia to New York.
How long does the local bus take to go from Philadelphia to New York?

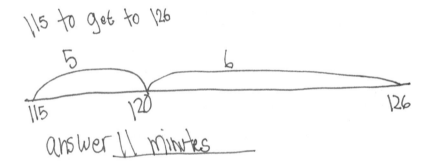

(c) What structures of the problem may account for the difficulty students experienced?
    (d) What next steps could Ms. Smith take to build students' ability to make sense of problems?

3. **Using Problem Structures to Select Tasks:** Figure 8.13 shows the strategy Myra used to arrive at a correct response to the bus task from the previous question.

Figure 8.13 Myra's response to the bus task

The express bus takes 126 minutes to go from Philadelphia to New York.

The local bus takes 115 minutes longer to go from Philadelphia to New York.

How long does the local bus take to go from Philadelphia to New York?

(a) On what level of the *OGAP Addition Progression* would Myra's response be found? What evidence in the work supports this analysis?

(b) Based on this analysis, select another task for Myra from the list below. What are your reasons for choosing that task?

Problem 1

Mike has 173 stickers in his collection
James has 129 stickers in his collection.
How many fewer stickers does James have than Mike?

Problem 2

Kim has 14 green grapes and 29 purple grapes.
Misha has 15 fewer grapes than Kim.
How many grapes does Misha have?

Problem 3

Anthony has 261 trading cards.
Gavin had 79 fewer trading cards than Anthony.
How many trading cards did Gavin have?

Problem 4

Angel was making a block building.
She wanted to use 20 blocks for her building.
She could use red or blue blocks.
Find as many combinations as you can.

4. **A Literacy Strategy for Sense Making with Problems:** Engaging students in sense-making strategies like the one outlined at the end of this chapter is an effective replacement to keyword-based approaches. To make sense of the text in mathematics problems, students need to find and synthesize important information and draw inferences. Consider the following problem, noticing that the question has been removed.

> **During Physical Education class students could choose one sport to play.**
> **25 students played in a soccer game,**
> **13 students played in a basketball game and**
> **16 played in a kickball game.**

(a) Make a list of mathematical questions students might ask about this situation.

(b) Try It! Look through your math program to select problems that can be used in this way. Select 2 or 3 problems that would give you a variety of questions. Remember to determine the problem type for each one selected so that you are providing opportunities with a variety of problem types.

Instructional Link

Use Table 8.2 and the ideas discussed in this chapter to help you consider ways in which your math instruction and math program provide students with opportunities to work with a variety of problem situations and structures.

1. What opportunities does your program provide for students to engage with problems that vary regarding the:
    (a) magnitude of numbers?
    (b) language and context?
    (c) units or quantities?
    (d) location of the unknown?

2. Are all problem situations identified in Table 8.2 at your grade level and in grades before included in your instruction or math program? Is there adequate exposure to the various problem situations across grades K–2 for students to be able to solve all 12 problem types of additive problem situations by the end of grade 2?

3. How could you modify given problems to differentiate for levels of developing understanding in students?

9
Developing Math Fact Fluency

Big Ideas

- Fluency with basic addition facts is critical for proficiency in mathematics and access to more complex mathematics concepts and tasks.
- Fact fluency is developed over time by building thinking and reasoning strategies on understanding of concepts, properties, and number relationships.
- Focusing on premature memorization of basic facts can be detrimental in the long term.
- Once fluency is attained, automaticity can be developed through both targeted and general practice.

There is little disagreement among mathematics educators on the importance of knowing basic addition facts; there is however a long history of debate over whether memorization and timed drill is the way to develop automaticity with these facts. This chapter considers the learning theories behind common instructional approaches to fact fluency, explores insights from mathematics education research, and unpacks strategies for building fact fluency on conceptual understanding supported by strategies and visual models.

The Importance of Fluency with Addition Facts

Teachers know from experience how important basic fact fluency in both addition and multiplication is for problem solving and more complex mathematics. Research confirms that a lack of fluency with basic facts leaves many students struggling, creating an ever-widening gap as they are less and less able to access advanced mathematics problems and concepts (Pellegrino & Goldman, 1987). Students who lack automaticity with basic facts may be less able to comprehend underlying mathematical concepts or access curriculum that increasingly emphasizes problem-solving and application tasks (Gersten et al., 2009). An example of the impact of a lack of fact fluency on problem solving can be seen in Kiria's approach in Figure 9.1.

Figure 9.1 Kiria's response. Kiria uses an inefficient early counting strategy to find a basic fact

A string was 56 inches long. Dylan cut some off. Now the string is 27 inches long. How much of the string did Dylan cut off?

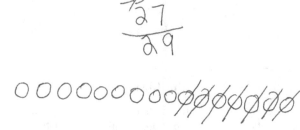

To solve this problem Kiria uses the Traditional US subtraction algorithm, an *Additive* strategy on the *OGAP Addition Progression,* but when she needs to subtract 7 from 16, she resorts to drawing 16 circles and crossing out 7, an *Early Counting* strategy. As discussed in Chapter 7, the Traditional US subtraction algorithm involves many steps to keep track of, alternating between regrouping and subtraction; having to stop and count to find a difference rather than drawing on a known addition fact, is both time-consuming and distracting. This is one reason why it is important for students to develop fluency with addition and subtraction facts before they are expected to use algorithms or efficient strategies for multi-digit numbers.

When recall of facts is fast and accurate, attention resources can be allocated to the more complex tasks for processing (Pellegrino & Goldman, 1987). In other words, using less working memory for the calculation of basic facts leaves more working memory for solving new or more complex problems (Bjorklund, Muir-Broaddus, & Schneider, 1990). In a study of children aged 7 to 12, Gray (1991) found that less successful or below average math students used more time-consuming counting methods for calculation while more successful students used recall or strategies that were based on numerical reasoning. As the math gets harder, this means that students who rely on counting are working harder as they face more challenging situations. Imagine trying to develop fluency with multiplication facts when students are still using counting strategies to derive addition facts. If students do not move away from inefficient counting as a strategy over time, they can become stuck in a cycle where doing the math becomes more and more difficult while they fall further behind (Gray & Tall, 1994).

Most teachers have observed how this drain on students' working memory impedes students' ability to keep up with the focus and flow of instruction (Forbinger & Fahsl, 2010). Sometimes students exert so much effort to derive a basic fact that they forget how the fact is related to the problem being solved. By the time they have labored through finding the fact, they have forgotten why they were looking for it.

Developing Fact Fluency: A Developmental Process

Fluency refers to the ability to recall a fact with relative ease, often through the use of some mental computation that results in an accurate and relatively efficient response. For example, when presented with 8 + 5, a student who can mentally decompose 5 into 3 and 2, combine 2 and 8 to make a 10 and know that 10 and 3 are 13, within 3 to 5 seconds, is fluent. The student is employing thinking strategies that make use of known facts and number relationships to quickly derive an unknown fact. In comparison, *automaticity* is effortless recall or retrieval from memory within about 3 seconds (Van de Walle et al., 2014). The distinction between fluency and automaticity is further clarified by McCallum (2012), a primary author of the CCSSM:

> [F]luently refers to how you do a calculation, whereas "know from memory" means being able to produce the answer when prompted without having to do a calculation. In the CCSSM, "fluent" means "fast and accurate."

Research shows that developing fluency is a developmental process (Baroody, 1985, 2006; Carpenter & Moser, 1984; Fuson, 1992; Henry & Brown, 2008). Baroody (2006) describes three phases in the process of developing fact fluency:

- Phase 1: *Counting Strategies* in which students count physical objects or verbally count to find "how many?";
- Phase 2: *Reasoning Strategies* in which students use strategies based on properties of operations and number relationships to derive unknown facts;
- Phase 3: *Mastery* in which a student can efficiently recall or produce answers.

Examine the *OGAP Addition and Subtraction Progressions*. Where do you see these phases reflected in the progressions?

Note that Phase 1 strategies are at the bottom of the progression, in the *Counting* level, while Phase 2 and 3 strategies are at the top in the *Additive* level. Importantly, the *Transitional* level, and in particular strategies that are anchored in visual models, are central to moving students from counting to additive strategies, from Phase 1 to Phase 2.

One may ask why not just focus on Phase 3 by having students memorize and practice single-digit addition facts? Research indicates prematurely focusing on memorization before students have sufficient experiences developing strategies for deriving facts can be detrimental to their developing number sense and conceptual understanding (Van de Walle et al., 2014). In addition, timed tests that focus on memorization can lead to math anxiety and overall aversion to the study of mathematics (Boaler, 2015).

Moreover, memorization is not always an effective approach. Human memory is associative, and this helps to explain why the memorization of isolated facts is difficult for many people. Information that is stored in long term memory through association with other ideas and understandings results in better retention and access to that information. Imagine the recall of a basic fact to be analogous to finding a specific book in a library. Without the organizing structures of a library that keep related books together, finding a specific book would require checking the title of every book in the library. However, recognizing that books on the same subject are grouped according to a specific structure allows one to go directly to the area where information on a certain subject can be found. Understanding how all the related books are organized on shelves and being able to use the landmark information found on each shelf further increases the speed with which the requested title can be found. When basic facts are stored in memory in association with conceptual understanding and strategies for deriving facts it is much easier and faster to find or produce a fact when needed.

A significant body of research supports the idea that focusing on number sense and reasoning strategies is more effective for learning basic facts than a focus on memorization and drill. (Baroody, 2006; Fuson, 1992; Henry & Brown, 2008). As shown in Figure 9.2, when students have built a strong foundation in conceptual understanding of the operations, the properties of operations, and number relationships, they can develop strategies for deriving facts that lead to fluency. Once students have developed strategies for fluently deriving facts, they can work on developing automaticity. An added benefit of this approach is that when students have difficulty memorizing or recalling facts, they have strategies and understanding to fall back upon. Rather than pulling out the wrong fact to solve a problem, they have the resources to figure it out.

Figure 9.2 An approach to automaticity that builds on fluency and conceptual understanding

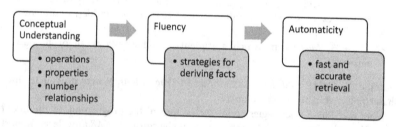

CCSSM Expectations for Fluency and Automaticity

The CCSSM reflects this developmental approach to fluency. Specifically, the standards reflect a progression of learning where automaticity develops out of fluency after significant attention has been given to building an understanding of important foundational concepts. The term fluently is used in the CCSSM at the endpoints of learning progressions that begin with conceptual understanding. One such endpoint is the second-grade standard for addition fact fluency. By the end of second grade, CCSSM expects that students should "know from memory all sums of two one-digit numbers." Progress towards knowing from memory can be traced back through a series of standards describing important concepts and strategies from kindergarten through second grade. This progression from the CCSSM, summarized below, mirrors the developmental progression from counting, to strategies for deriving facts, to recall of facts from memory.

- Kindergarten: Fluently add and subtract within 5 using counting strategies.
- Grade 1: Understand subtraction as a missing addend problem; relate counting to addition and subtraction.
- Grade 1: Fluently add and subtract within 10 using properties of operations and reasoning strategies such as counting on, making 10, decomposing a number, and the relationship between addition and subtraction.
- Grade 2: Fluently add and subtract within 20 and know from memory all sums of two one-digit numbers.

It is important to note that the CCSSM does not articulate fluency expectations for subtraction facts. Rather the attention is given to understanding the nature and application of the inverse relationship between addition and subtraction. The important foundational concepts represented in these standards—concepts of number composition, properties of operations, the relationship between addition and subtraction, number relationships, and strategies for addition and subtraction—are reflected in the *OGAP Additive Framework* and have been explored in detail throughout this book. If you have read all the prior chapters, you are in a good position to help your students develop fluency.

Fact Fluency: A Manageable Task

In additive reasoning, the basic facts are all the addition combinations of single digits 0 through 9 and the related subtraction facts. If one focuses on learning isolated facts, this means there are 100 different addition facts and 55 more subtraction facts to learn—a daunting feat! However, by making use of the ideas discussed throughout this book—developing understanding of properties and relationships, concepts, visual models, and strategies—developing fluency and automaticity with addition and subtraction facts can be a manageable task. This approach takes advantage of the associative nature of memory to build connections between facts.

Properties and Relationships

The understanding that addition and subtraction are inverse operations makes it easy to learn the subtraction facts, because every subtraction fact can be thought of as a related addition fact. For example, to subtract 13 − 5, a student can think, "what do I add to 5 to make 13?" This understanding is more than just learning to produce or memorize "fact families," or sets of related addition and subtraction equations with the same quantities. Discussing student solutions to problems situated in context, such as the one shown in Figure 9.3, can help students recognize the relationship between addition and subtraction. How might you use these solutions to focus on that understanding?

Figure 9.3 Student solutions can be used to focus on the inverse relationship between addition and subtraction

There are 12 cubes in all. How many cubes are hidden under the cup?

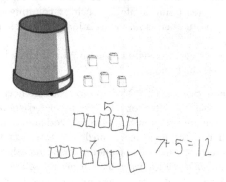

Nathan's solution

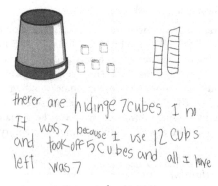

Brittany's solution

Note that Nathan finds the missing cubes by redrawing the five that are shown and then drawing 7 more to get to the total of 12. He also writes an addition sentence (7 + 5 = 12) to represent his solution. Brittany, on the other hand, starts with 12 cubes and then takes off 5 cubes to find the amount left (7). In discussing this solution, the teacher could help the class write a subtraction equation (12 − 5 = 7) and then talk about how the situation can be represented as addition (adding to) or subtraction (taking away). The relationship between addition and subtraction should be emphasized through context at every phase in the development of additive reasoning rather than as a procedural exercise of grouping related equations into "families."

 Go To Chapter 1 and Chapter 7 for more on developing understanding of the inverse relationship between addition and subtraction.

Another important property of addition is the understanding that adding zero to any number does not change the value. In mathematics, this is known as the *identity property of addition*. If this idea is not introduced directly, students may develop the incorrect generalization that "adding always makes bigger" (Van de Walle et al., 2014). Teachers can support the understanding of this concept through stories and situations where nothing is added or taken away from an existing quantity or from situations with two parts where one is empty. In modeling these situations, students can see when adding with concrete or visual models that adding no cubes to 5 cubes results in a total of 5 and conversely that 5 cubes plus no more cubes are also 5. Figure 9.4 illustrates how a story context can be used to help students connect with the meaning of the identity property of addition.

Figure 9.4 Additive situations to understand adding zero

Julie had two boxes of lollipops. There were 4 lollipops in one box and none in the other box. How many lollipops did she have?

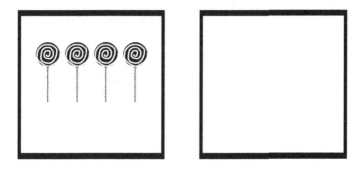

Understanding the *identity property of addition* along with the *commutative property*, the understanding that switching the order of addends does not change the result, reduces the number of addition facts to learn to only the 55 facts that are unshaded in Table 9.1. For example, if 3+4 equals 7 then the same digits added in another order would still equal 7, i.e., 3+4=7 and 4+3 also equals 7.

An understanding of commutativity can be seen in Jackson's work in Figure 9.5. Although he does not model the correct number of blocks in his drawing, the two equations he has written to represent the total number of blocks show evidence that he understands the commutative property. Giving students opportunities to interact with concrete and visual part-whole situations and allowing them to express the idea in their own words is important for developing understanding of the commutative property.

Table 9.1 Duplicate sums related to the commutative property

+	0	1	2	3	4	5	6	7	8	9
0	0	1	2	3	4	5	6	7	8	9
1	1	2	3	4	5	6	7	8	9	10
2	2	3	4	5	6	7	8	9	10	11
3	3	4	5	6	7	8	9	10	11	12
4	4	5	6	7	8	9	10	11	12	13
5	5	6	7	8	9	10	11	12	13	14
6	6	7	8	9	10	11	12	13	14	15
7	7	8	9	10	11	12	13	14	15	16
8	8	9	10	11	12	13	14	15	16	17
9	9	10	11	12	13	14	15	16	17	18

Figure 9.5 Jackson's response. Jackson's equations show an understanding of the commutative property of addition

Omar built 2 towers. How many blocks did he use altogether? Show how you know.

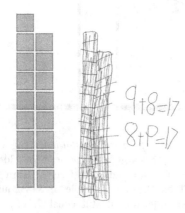

As discussed in Chapter 6, understanding commutativity also allows students to use more efficient strategies when counting on from a number, because they can count on from the larger number.

 Go To Chapter 6 for more on developing understanding of the commutative property.

Concepts, Strategies, and Visual Models

The big ideas of number discussed in Chapter 3 are also important and useful for developing fluency. *Hierarchical inclusion*, the concept that the value of a number contains all the previous values in the counting sequence, supports the fluent recall of plus 1 and plus 2 combinations. *Conceptual subitizing*, breaking a group into smaller subgroups that can be perceived visually, supports students' informal addition and subtraction strategies. For example, dot images like the ones shown in Figure 9.6 can be shown quickly to support the understanding and recall of "one more than" facts:

Figure 9.6 Dot images to highlight the concept of "one more than" through conceptual subitizing

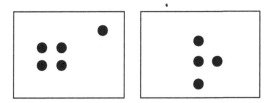

 Chapter 3 for more on hierarchical inclusion and conceptual subitizing.

Combinations that make 10 can be learned through subitizing activities with tens-frames, number paths, and bead strings, and Put Together/Take Apart problems that involve finding all the combinations of 10 (e.g., see Figure 8.8 in Chapter 8). Part–whole understanding, number composition, and anchoring numbers to 5 and 10 are also useful understandings to develop strategies for deriving facts. Students can draw upon their understanding of combinations of ten and the understanding of teen numbers as "a 10 and some more ones" to reduce the computational load when deriving sums over 10.

 Chapter 5 for more on visual models that can be used to help students learn the combinations of 10 and anchor numbers to 5 and 10.

Decomposing and recomposing the addends to find a combination of ten reduces the remaining calculation to putting a ten and some ones together to form a teen number. This combination of strategies based on an understanding of ten can be seen in Maisey's response in Figure 9.7 where she derives the sum of 9 and 4. To add 6, 3, and 4, she first uses the known fact of 6 + 3 = 9, then decomposes the 4 into 3 and 1, combines 9 and 1 to make a 10, and then knows that 10 and 3 more ones are the number 13.

Figure 9.7 Maisey's response shows evidence of decomposing an addend to make use of a known combination of 10 and understanding that teen numbers are composed of a 10 and some more ones

Shawn started to build a building with 6 blocks. Armondo added 3 blocks onto the building. Then Shawn put 4 more blocks on top. How many blocks did Shawn and Armando use for their building?

I know 6+3=9 and I put 1 of the 4 to the 9 which =10. I know 3+10=13

13

The kind of reasoning strategies shown in this response are important not only for fact fluency, but for later work with regrouping tens in algorithms, yet were found by Henry and Brown (2008) to be underrepresented in American classrooms. As shown on the *OGAP Addition and Subtraction Progressions*, the use of visual models like ten frames, ten strips, and ten-structured bead number lines, provides strong visual support as students develop strategies of making a 10 and anchoring to 10. Consider the pair of ten-strips in Figure 9.8 first discussed in Chapter 5.

Figure 9.8 How could these ten strips be used to find the sum of 9 + 8?

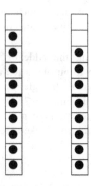

To visualize the "make a ten" strategy, one can mentally move dots from one strip to the other to fill in a ten. For example, moving a dot from the strip with 8 to the strip with 9 will result in the representation now showing the combination of 10 and 7, which is easy to recognize as 17. The same model can also be used to visualize how to make a 10 by adding 2 to the 8, resulting in 7 and 10. The visual supports in the ten-structured models discussed in Chapter 5 make them useful in supporting fact fluency. Repeated exposure to models that highlight 10 help students create a mental model for the quantity of 10 so that they can flexibly compose ten to derive basic facts with sums that are teen numbers.

 Go To Chapter 5 Visual Models to Support Additive Reasoning for more information on ten-structured models.

Since any sum over 10 can be derived by making a 10 and adding the remaining amount, nearly a third of the single-digit addition facts can be derived with this strategy (See Table 9.3).

Table 9.2 Combinations of 10 and sums over ten that can be derived with the "make a ten" strategy.

+	0	1	2	3	4	5	6	7	8	9
0										
1										10
2									10	11
3								10	11	12
4							10	11	12	13
5						10	11	12	13	14
6					10	11	12	13	14	15
7				10	11	12	13	14	15	16
8			10	11	12	13	14	15	16	17
9		10	11	12	13	14	15	16	17	18

Lastly, the set of facts known as doubles are often easily internalized and remembered. Once a student knows the doubles facts, they become another set of anchoring facts that can be used to efficiently derive near doubles. Combining what is known about plus 1 and minus 1 with the set of doubles facts allows for an additional 18 facts to be quickly accessed. Figure 9.9 shows how Will used his knowledge of 5 + 5 to find 5 + 4.

Figure 9.9 Will's strategy. Will uses a doubles fact to find a near double

If Daniel had 5 cupcakes, and he made 4 more, how many cupcakes would he have?

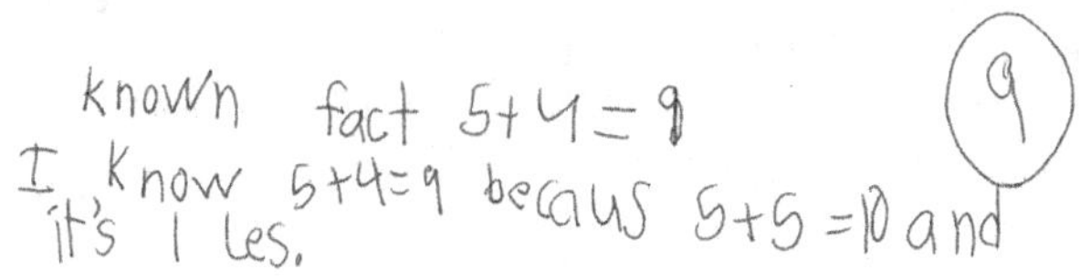

known fact 5+4 = 9
I know 5+4=9 becaus 5+5=10 and
it's 1 les. (9)

The set of facts that can be easily derived from doubles is illustrated in Table 9.3. For example, 4 + 3 can be derived as one less than 4 + 4 or one more than 3 + 3.

Table 9.3 Doubles and near doubles facts

+	0	1	2	3	4	5	6	7	8	9
0		1								
1	1	2	3							
2		3	4	5						
3			5	6	7					
4				7	8	9				
5					9	10	11			
6						11	12	13		
7							13	14	15	
8								15	16	17
9									17	

Figure 9.10 shows the same visual model from Figure 9.7 used to derive the sum from doubles. By matching up the equivalent dots on the two strips one can see that the sum is one more than 8 + 8. The same model can be used to see that the sum is one less than 9 + 9.

Figure 9.10 Using ten strips to derive facts from doubles

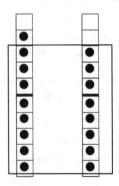

These strategies that anchor near doubles to doubles facts further exemplify the idea of developing strategies based on understanding and number relationships to develop fact fluency. In Figure 9.11 Charlie uses a known doubles fact to derive another fact, even though the needed fact was 3 less.

Figure 9.11 Charlie's response. Evidence of using known double fact 6 + 6 = 12 to find the sum of 6 + 3

There are 6 crayons in the bucket and 3 more on the table.

How many crayons are there altogether?

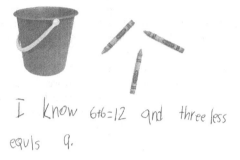

I know 6+6=12 and three less equls 9.

A student benefiting from instruction in strategies, concepts, relationships, and properties of addition built over the K–2 experience would be left with only two facts (5 + 3 and 6 + 3) that were not covered by any of the strategies or relationships previously discussed. These are not hard facts to learn, particularly if students are fluent with finding one and two more than any number. These facts are also easily derived from other known facts (e.g., using compensation, 5 + 3 is equivalent to 4+ 4).

Table 9.4 summarizes the strategies and properties discussed in this chapter for supporting the development of fluency and automaticity. Repeated and meaningful exposure to the strategies in the table builds fluency. As students become more fluent, they approach automaticity, which can be further supported by appropriate practice.

Table 9.4 Strategies supporting the development of fluency discussed in this chapter

Strategy	Used to derive	Examples
Identity property	Facts in which one addend is 0	4 + 0 = 4 0 + 6 = 6
Commutative property	Turn-around facts	5 + 3 = 8 3 + 5 = 8
1 more than, 2 more than	Facts with 1 or 2 as an addend	3 + 2 = 5 2 + 7 = 9 6 + 1 = 7 1 + 5 = 6
Doubles	Facts in which both addends are the same	3 + 3 = 6 8 + 8 = 16
Near doubles	Facts in which one addend is 1 away from the other	4 + 5 = (4 + 4) + 1 = 9 4 + 5 = (5 + 5) – 1 = 9
Compensation	Facts where the addends are 2 away from each other	6 + 8 = 7 + 7= 14
Combinations of 10	Facts that sum exactly 10	1 + 9 = 10 6 + 4 = 10
Make a 10	Facts whose sum is greater than 10	8 + 7 = 8 + (2 + 5) = (8 +2) + 5 = 10 + 5 = 15

From Fluency to Automaticity

Throughout this book we have described how instruction should be focused on developing conceptual understanding through subitizing, number relationships, relative magnitude, and visual models that support base-ten understanding. This instructional attention also benefits the development of math fact fluency by creating a network of number relationships that build on each other and can then be leveraged towards automaticity (Baroody, 1999; Davenport, Henry, Clements, & Sarama, 2019). Once students have developed strategies based on number relationships, practice can be used to strengthen associations and build automaticity, or effortless recall. Research has shown that engaging in practice before there is an association between the addition facts and their answers (e.g., when students are still using counting strategies) is not effective (Hasselbring, Goin, & Bransford, 1988). We discuss two kinds of practice: targeted practice to build fluency and automaticity and general practice to keep the associations that are developed at the forefront.

Targeted Fact Practice

Targeted practice involves focusing only on the specific facts with which the student is not yet automatic or fluent. The first step is to identify the facts a student knows or does not yet know.

Math Fact Interviews: This approach involves spending a few minutes conducting one-on-one interviews with students. First, prepare a set of flashcards representing all facts through 9 + 9 with each combination of addends present only once as shown in Figure 9.12. Try to vary the placement of the larger addend. The resulting set of 55 flashcards can then be presented in random order to determine fluent and automatic facts for individual students.

Figure 9.12 Streamline a set of fact cards for identifying student-specific practice sets

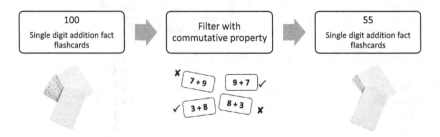

As the student responds to the flashcards, sort the cards into 3 piles as shown in Figure 9.13 based on the student's responses:

1. a pile for those the student accurately recalls within 3 seconds (automatic);
2. a pile for those the student can derive within 3 to 5 seconds using a strategy (fluent); and
3. a pile for those the student cannot figure out (does not know).

Figure 9.13 Sample piles and recording sheet from fact interview

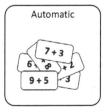 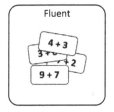 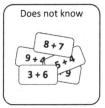

A sample student-specific recording sheet for recording the outcome of the student interview is shown in Figure 9.14. In the example, the interviewer has marked out facts that are automatic for the student and circled facts with which the student is fluent. The remaining unmarked facts represent the ones the student does not know. This interview process identifies a set of facts unique to the student that can then be used for individual targeted fact practice.

Figure 9.14 Sample recording sheet for math fact information

0+0 1+1 2+2 3+3 4+4 5+5 6+6 7+7 8+8 9+9 10+10

0+1 1+2 2+3 (3+4) 4+5 (5+6) 6+7 7+8 (8+9) 9+10

0+2 1+3 2+4 (3+5) 4+6 (5+7) (6+8) (7+9) 8+10

0+3 1+4 2+5 3+6 (4+7) (5+8) 6+9 7+10

0+4 1+5 2+6 3+7 (4+8) 5+9 6+10

0+5 1+6 (2+7) (3+8) 4+9 5+10

0+6 1+7 2+8 3+9 4+10

0+7 1+8 2+9 3+10

0+8 1+9 2+10

0+9 1+10

0+10

Flashcards: Practicing facts with flashcards where there is an addition fact on one side and the corresponding sum on the other side is a familiar approach. This format can be useful for moving from fluent to automatic with targeted facts. Students should create individualized, or personal flashcards in this format for facts found in their fluent pile during the math fact interview.

Additional opportunities are needed to help students learn strategies that can be used to derive the facts in their *does not know* pile from fact interviews. For these, students can be supported in creating individualized flashcards with both the target fact and hint—a related fact, visual model, or strategy selected by the student. These flashcards should have both the fact and the hint visible on the front of the card with the sum on the reverse. See examples in Figure 9.15. The hint serves as a launching point for deriving the unknown fact during practice activities. Take care not to impose a certain strategy, model, or relationship or imply that one is better than another. The goal is to support the student in selecting known information that is useful to them.

Figure 9.15 Nick's individualized flashcards for 6 + 7 and 7 + 8

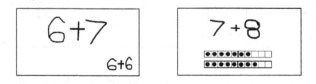

Targeted Practice Opportunities: Targeted practice opportunities should take place two to three times weekly. The number of facts with which a student needs to build automaticity will vary. It is valuable to limit the work to a manageable set of no more than 10 facts at a time. A student's practice set should include cards from both the fluent pile and the unknown pile to vary the difficulty. As students gain automaticity with facts in this set of 10, new facts that need attention can be added to the practice set. Individualized fact practice cards can be used in a variety of ways in and out of school for both independent and partner activities. One possible activity pairs students who then take turns quizzing each other on a small set of their individualized flashcards. As students continue to gain strategies and conceptual understanding through instructional activities, it is important to regularly check in with students to gather ongoing information about their progress towards fluency and which facts have become automatic.

General Fact Practice

General practice approaches, such as games and online resources that encourage reasoning and strategies based on important number concepts, can be more engaging for students than drill approaches. Games are fun and can build fluency by creating opportunities for repeated practice and strategic thinking. Games that involve student-to-student interaction also offer opportunities for students to learn strategies from each other (Forbinger & Fahsl, 2010). After students have played several rounds of a game, whole-class discussions can be used to highlight effective strategies, underlying relationships, or mathematical concepts. Both targeted fact practice and general fact practice should take place regularly in order to build fluency and automaticity.

Math games that can be effectively used to build fluency and automaticity are widely available. Published books of math games and routines, free online games and interactives, as well as many mathematics curricula, are robust sources of engaging general practice activities. When selecting general practice activities attention should be given to the structure of the game. Does the game simply reward students who already know their facts, or does it offer opportunities for students to reason, see relationships, or learn and develop new strategies?

Chapter Summary

- Addition fact fluency should be built upon conceptual understanding, properties of addition, and number relationships.
- Subtraction facts are learned through developing conceptual understanding of the inverse relationship between addition and subtraction.
- Focusing on drill and memorization activities before students have experiences developing strategies based on understanding can have a negative effect on students' number sense as well as their perceptions of mathematics and of themselves as learners of mathematics.
- Visual models such as dot images, ten frames, and ten strips can support students in developing strategies for deriving unknown facts from known facts.
- It is important to provide students with regular opportunities for two types of fact practice, targeted and general, to help them reach the goal of automaticity with the addition facts.

Looking Back

1. **Properties of Addition and Number Relationships:** Understanding number relationships and properties of addition are foundational to attaining fact fluency and must be integrated with an understanding of the inverse relationship between addition and subtraction from the earliest additive experiences. Consider the following task.

   Mark said 7 + 1 is equal to 1 + 7, so 7 − 1 is equal to 1 − 7. Is he right or Wrong? Show or explain how you know.

   (a) What evidence of student understanding might a teacher gather with this task?
   (b) Describe a couple of instructional activities that could be used to support students with misconceptions as they make sense of the task?

2. **Fluency with Addition Facts:** Fluency is a developmental process and takes a considerable amount of time to build. Baroody (2006) describes three phases in the process of developing fact fluency: Phase 1: Counting Strategies; Phase 2: Reasoning Strategies; Phase 3: Mastery.
   (a) Where are the phases suggested by Baroody reflected on the *OGAP Addition and Subtraction Progressions*?
   (b) How do *Transitional* strategies play a role in moving students from counting to additive strategies?

3. **Supporting Fact Fluency with Visual Model:** In Figure 9.16 Madelene uses a known fact to derive an unknown fact. How could you use a visual model to illustrate Madelene's response to her classmates?

Figure 9.16 Madelene's response using a known fact to derive an unknown fact

There are 6 crayons in the bucket and 3 more on the table.

How many crayons are there altogether?

6+3=9

I know that 5+3=8 So I know that 6+3=9.

Madelene's response

Instructional Link

Use the following questions to consider how your instruction and math program provide opportunities and support to learn the basic addition facts. Consider the math program for previous grades to determine what attention has been given to important concepts and strategies before your grade.

1. Does the instruction students receive for learning basic addition facts reflect the progression of understanding in Figure 9.2?
2. Are there enough opportunities for students to be exposed to and practice strategies based on commutativity, number relationships, base-ten understanding, and properties of operations?
3. Does the program provide guidance or methods for determining which basic facts each student knows and does not yet know? What steps would you need to take to provide personalized practice opportunities, both general and targeted?
4. How can ideas from this chapter be used to supplement the math program in any identified areas of need?

References

Baroody, A. J. (1984). Children's difficulties in subtraction: Some causes and questions. *Journal for Research in Mathematics Education, 15*(3), 203–213.

Baroody, A. J. (1985). Mastery of the basic number combinations: Internalization of relationships or facts? *Journal for Research in Mathematics Education, 16*(2), 83–98.

Baroody, A. J. (1999). Children's relational knowledge of addition and subtraction. *Cognition and Instruction, 17*(2), 137–175.

Baroody, A. J. (2006). Why children have difficulties mastering the basic number combinations and how to help them. *Teaching Children Mathematics, 13*(1), 22–31.

Baroody, A. J., Ginsburg, H. P., & Waxman, B. (1983). Children's use of mathematical structure. *Journal for Research in Mathematics Education, 14*(3), 156–168.

Bass, H. (2003). Computational fluency, algorithms, and mathematical proficiency: One mathematician's perspective. *Teaching Children Mathematics, 9*(6), 322–328.

Berch, D. B. (2005). Making sense of number sense: Implications for children with mathematical disabilities. *Journal of Learning Disabilities, 38*(4), 333–339.

Bjorklund, D. F., Muir-Broaddus, J. E., & Schneider, W. (1990). The role of knowledge in the development of strategies. In D. F. Bjorklund (Ed.), *Children's strategies: Contemporary views of cognitive development* (pp. 93–126). Hillsdale, NJ: Lawrence Erlbaum.

Black, P. J., & Wiliam, D. (2009). Developing the theory of formative assessment. *Educational Assessment, Evaluation and Accountability, 21*(1), 5–31.

Boaler, J. (2015). Fluency without fear: Research evidence on the best ways to learn math facts. *Reflections, 40*(2), 7–12.

Bray, W. S., & Maldonado, L. A. (2018). Foster fact fluency with number-string discussions. *Teaching Children Mathematics, 25*(2), 86–93.

Carpenter, T. P., & Fennema, E. (1992). Cognitively guided instruction: Building on the knowledge of students and teachers. *International Journal of Research in Education, 17*, 457–470.

Carpenter, T. P., Fennema, E., Franke, M. L., Levi, L., & Empson, S. B. (2015). *Children's mathematics: Cognitively guided instruction* (2nd ed.). Portsmouth, NH: Heinemann.

Carpenter, T. P., Fennema, E., Peterson, P. L., Chiang, C.-P., & Loef, M. (1989). Using knowledge of children's mathematics thinking in classroom teaching: An experimental study. *American Educational Research Journal, 26*(4), 499–531.

Carpenter, T. P., Franke, M. L., Jacobs, V. R., Fennema, E., & Empson, S. B. (1998). A longitudinal study of invention and understanding in children's multidigit addition and subtraction. *Journal for Research in Mathematics Education, 29*(1), 3–20.

Carpenter, T. P., Franke, M. L., Johnson, N. C., Turrou, A. C., & Wager, A. (2017). *Young children's mathematics: Cognitively guided instruction in early childhood education*. Portsmouth, NH: Heinemann.

Carpenter, T. P., Franke, M. L., & Levi, L. (2003). *Thinking mathematically: Integrating arithmetic and algebra in elementary school*. Portsmouth, NH: Heinemann.

Carpenter, T. P., & Moser, J. M. (1984). The acquisition of addition and subtraction concepts in grades one through three. *Journal for Research in Mathematics Education, 15*(3), 179–202.

Carraher, T. N., Carraher, D. W., & Schliemann, A. D. (1987). Written and oral mathematics. *Journal for Research in Mathematics Education, 18*, 83–97.

Ching, B. H. H., & Nunes, T. (2017). The importance of additive reasoning in children's mathematical achievement: A longitudinal study. *Journal of Educational Psychology, 109*(4), 477.

Clement, L. L., & Bernhard, J. Z. (2005). A problem-solving alternative to using key words. *Mathematics Teaching in the Middle School, 10*(7), 360–365.

Clements, D. H. (2000). 'Concrete' manipulatives, concrete ideas. *Contemporary Issues in Early Childhood, 1*(1), 45–60.

Clements, D. H., & Sarama, J. (2014). *Learning and teaching early math: The learning trajectories approach.* Abingdon, Oxon: Routledge.

Clements, D. H., Sarama, J., Baroody, A. J., & Joswick, C. (2020). Efficacy of a learning trajectory approach compared to a teach-to-target approach for addition and subtraction. *ZDM*, 1–12. doi:10.1007/s11858-019-01122-z

Clements, D. H., Sarama, J., Spitler, M. E., Lange, A. A., & Wolfe, C. B. (2011). Mathematics learned by young children in an intervention based on learning trajectories: A large-scale cluster randomized trial. *Journal for Research in Mathematics Education, 42*(2), 127–166.

Clements, D. H., Sarama, J., Wolfe, C. B., & Spitler, M. E. (2013). Longitudinal evaluation of a scale-up model for teaching mathematics with trajectories and technologies: Persistence of effects in the third year. *American Educational Research Journal, 50*(4), 812–850.

Cobb, P. (1995). Cultural tools and mathematical learning: A case study. *Journal for Research in Mathematics Education, 26*(4), 362–385.

Cobb, P., & Wheatley, G. (1988). Children's initial understandings of ten. *Focus on Learning Problems in Mathematics, 10*(3), 1–27.

Common Core Standards Writing Team. (2019). *Progressions for the common core state standards for mathematics (Draft February 7, 2019).* Tucson, AZ: Institute for Mathematics and Education, University of Arizona.

Davenport, L. R., Henry, C. S., Clements, D. H., & Sarama, J. (2019). *No more math fact frenzy.* Portsmouth, NH: Heinemann.

Denton, K., & West, J. (2002). *Children's reading and mathematics achievement in kindergarten and first grade* (National Center for Education Statistics Report No. NCES 2002-125). Washington, DC: Ed Pubs.

Drake, J. M., & Barlow, A. T. (2008). Assessing students' levels of understanding multiplication through problem writing. *Teaching Children Mathematics, 14*(5), 272–277.

Duncan, G. J., Dowsett, C. J., Claessens, A., Magnuson, K., Huston, A. C., Klebanov, P., … Japel, C. (2007). School readiness and later achievement. *Developmental Psychology, 43*(6), 1428–1446.

Ebby, C. B. (2005). The powers and pitfalls of algorithmic knowledge: A case study. *Journal of Mathematical Behavior, 24*, 73–87.

Ebby, C. B., Hulbert, E. T., & Fletcher, N. (2019). What can we learn from correct answers? *Teaching Children Mathematics, 25*(6), 346–353.

Forbinger, L., & Fahsl, A. J. (2010). Differentiating practice to help students master basic facts. In D. Y. White & J. S. Spitzer (Eds.), *Responding to diversity: Grades pre K–5* (pp. 7–22). Reston, VA: NCTM.

Fosnot, C. T., & Dolk, M. L. (2001). *Young mathematicians at work: Constructing number, addition and subtraction.* Portsmouth, NH: Heinemann.

Franke, M. L., Kazemi, E., & Turrou, A. C. (2018). *Choral counting and counting collections: Transforming the prek–5 math classroom.* Portsmouth, NH: Stenhouse.

Fuson, K. C. (1984). More complexities in subtraction. *Journal for Research in Mathematics Education, 15*(3), 214–225.

Fuson, K. C. (1986). Teaching children to subtract by counting up. *Journal for Research in Mathematics Education, 17*(3), 172–189.

Fuson, K. C. (1988). *Children's counting and concepts of number.* New York: Springer Verlag.

Fuson, K. C. (1992). Research on whole number addition and subtraction. In D. A. Grouws (Ed.), *Handbook of research on mathematics teaching and learning* (pp. 243–275). New York: Macmillan Publishing Company.

Fuson, K. C., & Beckmann, S. (2012). Standard algorithms in the common core state standards. *NCSM Journal, 14*(2), 14–30.

Fuson, K. C., & Briars, D. J. (1990). Using a base-ten blocks learning/teaching approach for first- and second-grade place-value and multidigit addition and subtraction. *Journal for Research in Mathematics Education, 21*, 180–206.

Fuson, K. C., Carroll, W. M., & Landis, J. (1996). Levels in conceptualizing and solving addition and subtraction compare word problems. *Cognition and Instruction, 14*(3), 345–371.

Fuson, K. C., Kalchman, M., & Bransford, J. D. (2005). Mathematical understanding: An introduction. In M. S. Donovan & J. Bransford (Eds.), *How students learn mathematics in the classroom* (pp. 217–256). Washington, DC: National Research Council.

Fuson, K. C., Wearne, D., Hiebert, J., Human, P., Murray, H., Olivier, A., ... Fennema, E. (1997). Children's conceptual structures for multidigit numbers and methods of multidigit addition and subtraction. *Journal for Research in Mathematics Education*, *28*, 130–162.

Gambrell, L. B., Koskinen, P. S., & Kapinus, B. A. (1991). Retelling and the reading comprehension of proficient and less-proficient readers. *The Journal of Educational Research*, *84*(6), 356–362.

Gelman, R., & Gallistel, R. C. (1978). *The child's understanding of number*. Cambridge, MA: Harvard University Press.

Gersten, R., Chard, D. J., Jayanthi, M., Baker, S. K., Morphy, P., & Flojo, J. (2009). Mathematics instruction for students with learning disabilities: A meta-analysis of instructional components. *Review of Educational Research*, *79*(3), 1202–1242.

Gravemeijer, K. (1999). How emergent models may foster the constitution of formal mathematics. *Mathematical Thinking and Learning*, *1*(2), 155–177.

Gravemeijer, K. P. E., & van Galen, F. H. J. (2003). Facts and algorithms as products of students' own mathematical activity. In J. Kilpatrick, W. G. Martin, & D. Schifter (Eds.), *A research companion to principles and standards for school mathematics* (pp. 114–122). Reston: VA: NCTM.

Gray, E., & Tall, D. (1994). Duality, ambiguity, and flexibility: A "proceptual" view of simple arithmetic. *Journal for Research in Mathematics Education*, *25*(2), 116–140.

Gray, E. M. (1991). An analysis of diverging approaches to simple arithmetic: Preference and its consequences. *Educational Studies in Mathematics*, *22*(6), 551–574.

Hasselbring, T. S., Goin, L. I., & Bransford, J. D. (1988). Developing math automatically in learning handicapped children: The role of computerized drill and practice. *Focus on Exceptional Children*, *20*(6), 1–7.

Hegarty, M., Mayer, R. E., & Monk, C. A. (1995). Comprehension of arithmetic word problems: A comparison of successful and unsuccessful problem solvers. *Journal of Educational Psychology*, *87*(1), 18.

Henry, V. J., & Brown, R. S. (2008). First-grade basic facts: An investigation into teaching and learning of an accelerated, high-demand memorization standard. *Journal for Research in Mathematics Education*, *39*(2), 153–183.

Hiebert, J., & Wearne, D. (1996). Instruction, understanding, and skill in multidigit addition and subtraction. *Cognition and Instruction*, *14*(3), 251–283.

Howe, R., & Epp, S. S. (2008). Taking place value seriously: Arithmetic, estimation and algebra. *Resources for RMET (Preparing Mathematicians to Educate Teachers)*. Retrieved from www.maa.org/sites/default/files/pdf/pmet/resources/PVHoweEpp-Nov2008.pdf

Hulbert, E. T., Petit, M. M., Ebby, C. B., Cunningham, E. P., & Laird, R. E. (2017). *A focus on multiplication and division: Bringing research to the classroom*. New York: Taylor & Francis.

Kalchman, M., Moss, J., & Case, R. (2001). Psychological models for the development of mathematical understanding: Rational numbers and functions. In S.M. Carver & D. Klahr (Eds.), *Cognition and instruction: Twenty-five years of progress* (pp. 1–38). Mahwah, NJ: Lawrence Elbaum.

Kamii, C. (1982). Encouraging thinking in mathematics. *Phi Delta Kappan*, *64*(4), 247–251.

Kamii, C., & Dominick, A. (1998). The harmful effects of algorithms in grades 1–4. *The Teaching and Learning of Algorithms in School Mathematics*, *19*, 130–140.

Kamii, C., & Joseph, L. L. (2004). *Young children continue to reinvent arithmetic–2nd grade: Implications of Piaget's theory*. New York, NY: Teachers College Press.

Kamii, C. K. (1985). *Young children reinvent arithmetic: Implications of Piaget's theory*. New York, NY: Teachers College Press.

Kamii, C. K. (1989). *Young children continue to reinvent arithmetic, second grade*. New York, NY: Teachers College Press.

Kilpatrick, J., Swafford, J., & Findell, B., (Eds.). (2001). *Adding it up: Helping children learn mathematics* National Research Council (Ed.). Washington, DC: National Academy Press.

Klein, A. S., Beishuizen, M., & Treffers, A. (1998). The empty number line in Dutch second grades: Realistic versus gradual program design. *Journal for Research in Mathematics Education*, *29*, 443–464.

Krebs, G., Squire, S., & Bryant, P. (2003). Children's understanding of the additive composition of number and of the decimal structure: What is the relationship? *International Journal of Educational Research*, *39*(7), 677–694.

Lambert, R., Imm, K., & Williams, D. A. (2017). Number strings: Daily computational fluency. *Teaching Children Mathematics, 24*(1), 48–55.

Martins-Mourão, A., & Cowan, R. (1998). The emergence of additive composition of number. *Educational Psychology, 18*(4), 377–389.

McCallum, B. (2012, April 26). General questions about the mathematics standards. Retrieved from www.math.arizona.edu/~ime/2011-12/blogdiscussion_2012_05_26.pdf

Metsisto, D. (2005). Reading in the mathematics classroom. In J. M. Kenney, E. Hancewicz, L. Heuer, D. Metsisto, & C. L. Tuttle (Eds.), *Literacy strategies for improving mathematics instruction* (pp. 9–23). Alexandria, VA: ASCD.

Morrow, L. M. (1985). Retelling stories: A strategy for improving young children's comprehension, concept of story structure, and oral language complexity. *The Elementary School Journal, 85*(5), 647–661.

National Council of Teachers of Mathematics. (2000). *Principles and standards for school mathematics.* Reston, VA: NCTM.

National Council of Teachers of Mathematics. (2014a). *Principles to actions: Ensuring mathematical success for all.* Reston, VA: NCTM.

National Council of Teachers of Mathematics. (2014b). *Procedural fluency in Mathematics: A position of the National Council of Teachers of Mathematics.* Retrieved from www.nctm.org/Standards-and-Positions/Position-Statements/Procedural-Fluency-in-Mathematics/

National Governors Association & Council of Chief State School Officers. (2010). *Common core state standards for mathematics.* Washington, DC: Author. Retrieved from www.corestandards.org/Math/

National Mathematics Advisory Panel. (2008). *Foundations for success: The final report of the National Mathematics Advisory Panel.* Washington, DC: U.S. Department of Education.

National Research Council. (2009). *Mathematics learning in early childhood: Paths toward excellence and equity.* Washington, DC: National Academies Press.

Nunes, T. (1992). Ethnomathematics and everyday cognition. In D. A. Grouws (Ed.), *Handbook of research on mathematics teaching and learning* (pp. 557–574). New York: Macmillan.

Nunes, T., & Bryant, P. (1996). *Children doing mathematics.* Oxford: Blackwell Publishers.

Pape, S. J., & Tchoshanov, M. A. (2001). The role of representation (s) in developing mathematical understanding. *Theory into Practice, 40*(2), 118–127.

Pellegrino, J. W., & Goldman, S. R. (1987). Information processing and elementary mathematics. *Journal of Learning Disabilities, 20*(1), 23–32.

Piaget, J. (1965). *The child's conception of number.* New York: Norton.

Ross, S. H. (1989). Parts, wholes, and place value: A developmental view. *Arithmetic Teacher, 36* (6), 47–51.

Russell, S. J. (2000). Developing computational fluency with whole numbers. *Teaching Children Mathematics, 7*(3), 154–158.

Sarama, J., & Clements, D. H. (2009). *Early childhood mathematics education research: Learning trajectories for young children.* New York: Routledge.

Saxe, G. B. (1988). Candy selling and math learning. *Educational Researcher, 17*(6), 14–21.

Schwerdtfeger, J. K., & Chan, A. (2007). Counting collections. *Teaching Children Mathematics, 13*(7), 356–361.

Shumway, J. (2011). *Number sense routines, grades K–3: Building numerical literacy every day in grades K–3.* Portsmouth, NH: Stenhouse.

Sinclair, A., Mello, D., & Siegrist, F. (1988). La notation numérique chez l'enfant. In H. Sinclair (Ed.), *La production de notations chez le jeune enfant: langage, nombre, rythmes et mélodies* (pp. 71–98). Paris: Presses Universitaires de France.

Sousa, D. A. (2008). *How the brain learns mathematics.* Thousand Oaks, CA: Corwin Press.

Sowder, L. (1988). Children's solutions of story problems. *The Journal of Mathematical Behavior, 7*(3), 227–238.

Sowell, E. J. (1989). Effects of manipulative materials in mathematics instruction. *Journal for Research in Mathematics Education, 20*(5), 498–505.

Stein, M. K., & Smith, M. (2018). *Five practices for orchestrating productive mathematics discussions.* Reston, VA: National Council of Teachers of Mathematics.

Supovitz, J. A., Ebby, C. B., Remillard, J., & Nathenson, R. A. (2018). *Experimental impacts of the ongoing assessment project on teachers and students* (CPRE Research Report# RR 2018-1). Philadelphia, PA: Consortium for Policy Research in Education.

Teacher Education by Design. (n.d.). *Number strings*. Mathematics Instructional Activities. Retrieved from https://tedd.org/number-strings/

Van de Walle, J., Karp, K. S., & Bay-Williams, J. M. (2014). *Elementary and middle school mathematics: Teaching developmentally* (8th international ed.). Essex, UK: Pearson.

Verschaffel, L., Greer, B., & DeCorte, E. (2007). Second handbook of research on mathematics teaching and learning. *Whole Number Concepts and Operations, 1,* 557–628.

Wilson, P. H., Sztajn, P., Edgington, C., & Myers, M. (2015). Teachers' uses of a learning trajectory in student-centered instructional practices. *Journal of Teacher Education, 66*(3), 227–244.

About the Authors

Caroline B. Ebby is a senior researcher at the Consortium for Policy Research in Education (CPRE) and an adjunct associate professor at the Graduate School of Education at the University of Pennsylvania. Caroline is also a member of the OGAP National Professional Development Team.

Elizabeth T. Hulbert is the managing partner of the Ongoing Assessment Project (OGAP) and oversees the OGAP National Professional Development Team. Beth previously worked in public education for many years as a classroom teacher and district mathematics coach.

Rachel M. Broadhead is site director for the Alabama Math, Science, and Technology Initiative (AMSTI) at the University of South Alabama Regional Site and leads a team of content specialists and coaches. Rachel is also a member of the OGAP National Professional Development Team.

Index

Page locators in *italics* and **bold** refer to figures and tables, respectively.

Printed in the United States
By Bookmasters